MICROMECHANICS OF FLOW IN SOLIDS

MICROMECHANICS OF FLOW IN SOLIDS

JOHN J. GILMAN

DIRECTOR, MATERIALS RESEARCH CENTER
ALLIED CHEMICAL CORPORATION

McGRAW-HILL BOOK COMPANY

New York, St. Louis, San Francisco, London
Sydney, Toronto, Mexico, Panama

Library of Congress Catalog Card Number 70-77956

23227

1 2 3 4 5 6 7 8 9 0 M A M M 7 6 5 4 3 2 1 0 6 9

Dedicated
to
ALEXANDER FALK GILMAN

PREFACE

This book is largely an exposition of a particular viewpoint regarding the physical nature and description of how solids flow when they are put under stress. Although thousands of valuable interactions between the author and his colleagues, both past and present, lie in the background, this book is not an attempt to meld various viewpoints. Rather, it represents a personalized selection of topics and their treatments. The justification of this approach to the subject is that it leads to a coherence that other approaches lack. The glue that binds the rambling structure of the subject together is the explicit introduction of, and emphasis on, time as a parameter. This not only provides an internal unity but also unites the physics of flow with the rest of solid-state physics.

Very many students have contributed to the content of this book through their puzzled responses to initial drafts and critical comments. Especially valuable has been the careful reading and checking by Dr. Hua-ching Tong.

John J. Gilman

CONTENTS

1

INTRODUCTION

My purpose is to present a unified description of the nature of plastic flow in solid matter. In its details plastic flow is enormously complicated because a plastic material is really a combination of solid, liquid, and gaseous elements. This will become more clear later on in the text, but the point to be made here is that a plastic material not only inherits the complications possessed by its several parts but adds some structural complexities of its own. A unified view is only possible, then, if many of the details are not considered. Thus the treatment to be presented is by no means comprehensive and is elementary.

The essence of a solid is its resistance to deformation, in other words, its strength. The superior ability of solids in this respect, and especially their ability to resist shearing forces, is what causes people to distinguish them from liquids and gases. The strength property has its maximum intensity when a solid is crystalline, because of the tight packing of atoms in crystals. This fact, together with the regular and therefore relatively simple structures of crystals, makes a knowledge of their strength the basis of understanding strength in noncrystalline solids.

Most of the solids encountered in nature and technology are either noncrystalline or polycrystalline aggregates, but there is every reason to believe that the same kinds of forces (and no essentially new ones) act in them as in crystals. However, their complex structures defy precise description, and so their properties cannot be related directly to interatomic forces and instead are described in terms of empirical measurements designed to rank various materials with respect to a property of particular interest.

Three broad characteristics come to mind when the meaning of the "strength" of a solid is considered. First, one thinks of its resilience, or elasticity, when it is subjected to transient forces; that is, how strongly does it attempt to regain its original shape if a force is applied to it for a time and then removed? Second, its resistance to flow comes to mind. If a sustained force is applied, how rapidly will the material undergo a permanent change in shape? Or the corollary, how much applied force is required to obtain a particular rate of flow? Third, solids that are pulled by enough tension can be broken into pieces. How much applied force is required to do this for a given solid?

The three characteristics mentioned above vary enormously in magnitude from one material to another. Thus elasticity, as measured by Young's modulus, varies over at least three orders of magnitude from about 10^{10} to 10^{13} dyn/cm². The flow resistance varies even more, stresses from about 10^5 to 10^{12} dyn/cm² being required to produce an observable flow rate. Also, measured flow rates extend over at least twelve orders of magnitude, and if rates inferred from geologic observations are included, the range is extended by several additional orders. Fracture resistance is the resistance of a material to crack propagation and is measured in terms of the fracture surface energy which may be as little as 10 dyn/cm or as much as 10^8 dyn/cm—a variation of seven orders of magnitude.

Since strength has such variable aspects, it should not be surprising that its study presents a large complex of problems. Only the considerable persistence of research workers in the field and the transcendent technological importance of the subject have led to progress in identifying some of its universal features. This has continually improved the physical and quantitative description that can be given of strength. However, strength depends on the collective response of a crystal to applied stresses, so its description is far more complex than that of a phenomenon like electrical conduction. Another way of saying this is that strength depends on the behavior of aggregates of particles (elasticity), surfaces composed of particles (fracture), or lines of particles (flow via dislocations), and not simply the behavior of single particles. Because of these complexities, strength phenomena have a rich texture but are difficult to describe in terms of simple numbers, so that one commonly resorts to graphic display.

The external mechanical variables that are of interest are the applied tractions, the shape of the material, and the time. Internal variables are the internal stress distribution (whose average must be equal and opposite to the applied tractions), the temperature, and the internal structure. The isothermal mechanical state of a crystal is then a point in a four-dimensional stress, shape, time, and structure space. However, stress requires a fourth-rank tensor for its description, and structure requires much more, and so it is apparent that only the simplest of situations can be discussed in detail, and even then we shall see that a statistical approach is required. There are some special cases when the situation is simplified, allowing a more detailed numerical description. Purely elastic behavior is the prime example, where time and structural changes do not play a role, so that stress and shape have a simple correlation through Hooke's law.

The mechanical behavior of a crystal or aggregate is most compactly described by means of a set of stress-elongation curves as shown schematically in Fig. 1.1. Here the line OPQ shows the general shape of a stress-strain

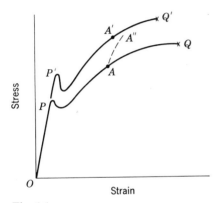

Fig. 1.1

curve where OP is the elastic portion and plastic flow occurs along PQ, followed by fracture at Q. A single curve is inadequate, however, because time is too important to be neglected. The line $OP'Q'$ shows schematically what happens if the time scale is changed by increasing the rate of elongation. The maximum that appears shortly after observable flow begins and is followed by a region of unstable (negative slope) flow is not always present, but is a typical feature.

The description is further complicated by the fact that structure has not been included as a parameter. Therefore, if the rate of elongation is suddenly changed from $\dot{\epsilon}_1$ to $\dot{\epsilon}_2$ at some point such as A in Fig. 1.1, the stress will not (in general) simply change from A on curve PQ to A' on curve $P'Q'$. Instead it may follow some other curve such as AA'', depending on what detailed structure is present at A.

Mechanical behavior may also be described using creep curves which show how the shape changes with time at constant stress, and stress-relaxation curves are sometimes used to show how the stress changes with time after a particular shape has been suddenly imposed. In all cases, the internal structure must be known in order to predict what will happen next at any particular point.

The central problem of the science of strength has been to learn how to describe mechanical behavior in terms of the fundamental properties of matter so that the shapes of stress-strain, strain-time, or stress-time curves and their dependence on sustained or sudden changes of loading conditions can be predicted from a small number of experimental relationships.

The crystalline solid state has played a central role in the learning process for theoretical, experimental, and technological reasons. Three-dimensional periodicity gives a crystal structural simplicity that greatly aids theoretical calculations and limits the number of mechanical mechanisms to a few possibilities. Crystals also have important advantages as experimental materials because they tend to be uniform and have been produced in recent years with a very high degree of perfection. Because of impurity rejection during solidification, crystals have a natural tendency toward purity which simplifies their internal structures. Furthermore, localized defects such as dislocations that appear during plastic deformation are strong perturbations in the structure of a good crystal, allowing the defects in them to be observed with various kinds of microscopes. This is much more difficult to do in a glassy structure, where a broad spectrum of defects tends to overwhelm a particular species of interest.

The materials that are used by engineers to withstand high mechanical stresses are almost always aggregates of crystals. This is true of metals, ceramics (except glasses), and high-strength polymers. The reason is that the lowest possible energy state is that in which the atoms or molecules of a solid are neatly and densely arranged on some simple crystal lattice. When such a solid is subjected to stresses that tend to make it deform or break, it endeavors to maintain its low-energy crystalline structure, and this gives a special character to its mechanical behavior. Also, it is the basis for the important role played by dislocations in structural materials, because dislocations minimize the volume of severely strained material, leaving a maximum of remaining crystallinity.

At the very centers of dislocation lines, however, the normal crystal structure is destroyed, and so the local region approaches a liquidlike state. Thus a dislocated solid can be thought (roughly) to consist of slender liquid channels snaking about inside a solid matrix. But if the solid is at a finite temperature, or if it contains mobile charge carriers, then quasi-particles (electrons, holes, phonons, etc.) dart about inside and interact with the liquidlike channels (dislocations). This is indeed a complex situation! Nevertheless, a relatively simple description of it can be given that allows vast quantities of data to be correlated and quantitative predictions to be made. Such a description is the focus of this book. Its spirit is statistical and also dynamical, that is, micromechanical.

In the future one can expect to see increasing technological use of individual crystals as stress-resisting devices, in response to a demand for high reliability and a maximum of strength. This trend will terminate a line of movement toward ever more regular internal structure to achieve high strength and reliability. The line began with the use of natural materials like stone and wood that were relatively weak and unreliable. It passed to materials like brick, glass, and steel that have complex structures in detail but are more homogeneous (less deviation of any element from the mean element) than natural materials. Then came a movement toward structural simplification by purification of these materials, resulting in such things as highly pure and dense ceramics, pure inorganic glasses and synthetic organic ones, and vacuum-melted special steels. Currently there is a tendency toward structural simplification through the elimination of crystal boundaries, which means toward the use of individual crystals as structural parts.

2

REVIEW OF
ELASTIC BEHAVIOR

Precision in the theory of elasticity began with the paper of Robert Hooke (1678), who stretched wires with known weights, bent spring-like beams, and twisted wires with known torques. After measuring their deflections, he concluded that "It is very evident that the Rule (or Law) of Nature in every springing body is, that the force (or power) thereof to restore itself to its natural position is always proportionate to the distance (or space) it is removed therefrom, whether it be by rarefaction (the separation of its parts the one from the other), or by a condensation (or crowding of its parts nearer together)." The linear relation that this expresses between force and deformation is now known as *Hooke's law* (Andrade, 1950).† It forms the foundation for the mechanics of elastic materials.

The usefulness of Hooke's law stems from the very high resistance to deflections that many solids possess. Because of this high resistance, quite large applied stresses (10^3 to 10^4 times atmospheric pressure) cause only small strains. Therefore, the law is obeyed for a large range of stresses. For strong crystals, it may apply for stresses as large as 10^5 atm.

In order to develop the theory, it is necessary to make some definitions. These can be appreciated most easily in the one-dimensional case. Although this case can be misleading because it does not include torsion and flexure, it nevertheless is a good place to start a review.

2.1 ONE-DIMENSIONAL ELASTICITY

Consider a one-dimensional rod that is fixed at one end and subjected to a force F acting at the other end (x direction). An element of the

† This refers to the work by Andrade, written in 1950, as listed in the References at the end of the chapter. This system of references will be used throughout the book.

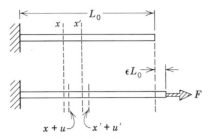

Fig. 2.1 Stretching of a one-dimensional elastic bar.

rod is defined by two reference positions x and x'. When the force is applied, stretching of the rod displaces these reference positions by amounts u and u', respectively. In general, *displacements* take place that are defined by some function $u(x)$. An elongation e of the rod occurs that is measured by the ratio of the stretched length L to the unstretched length L_0; that is,

$$e = \frac{L}{L_0} = \frac{(x' + u') - (x - u)}{x' - x} = 1 + \frac{u' - u}{x' - x}$$
$$= 1 + \frac{du}{dx}$$

(2.1)

The *strain* ϵ in the stretched rod is defined as the gradient of the displacement:

$$\epsilon = \frac{du}{dx} = e - 1$$

(2.2)

If the displacements change with time in the rod, a *particle velocity* $= du/dt$ may exist as well as *particle accelerations* $= d^2u/dt^2$. These are distinct from any wave velocities or accelerations that may also exist (see below).

The force on the rod per unit area A defines the *stress* $\sigma = F/A$ in it. The strain is proportional to this stress according to Hooke's law,

$$\epsilon = \frac{\sigma}{M}$$

(2.3)

and the proportionality constant is called *Young's modulus*.

Since elastic stretching is a reversible process (when done slowly), the strain energy that is stored in the rod equals the elastic work W done in stretching it (or compressing it). Since work equals force times displacement (at $x = L$ where the force is applied),

$$W = \int_0^u f \, du = \frac{L}{MA} \int_0^F f \, df = \frac{\sigma}{2M} AL$$

and the work per unit volume is the *strain energy density U:*

$$U = \frac{\sigma\epsilon}{2} = \frac{\sigma^2}{2M} = \frac{M\epsilon^2}{2} \qquad (2.4)$$

The *equation of motion* for an elastic rod is obtained by considering the force equilibrium of a mass element within it. Such an element is sketched in Fig. 2.2 being acted upon by stress forces at either end and an inertial

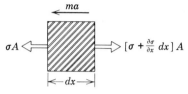

Fig. 2.2 Forces that act on a mass element in an elastic bar.

force determined by the mass times the acceleration,

$$ma = \rho A \, dx \, \frac{\partial^2 u}{\partial t^2}$$

where ρ is the mass density. The force equilibrium equation is

$$\Sigma F_x = \left(\sigma + \frac{\partial\sigma}{\partial x} \, dx \right) A - \sigma A - \rho A \, dx \, \frac{\partial^2 u}{\partial t^2} = 0$$

so the stress gradient equals the mass density times the acceleration:

$$\frac{\partial\sigma}{\partial x} = \rho \, \frac{\partial^2 u}{\partial t^2} \qquad (2.5)$$

If there is no acceleration, the stress gradient must be zero in order to have static equilibrium:

$$\frac{\partial\sigma}{\partial x} = 0$$

These equations are usually made to be homogeneous by noting that $\sigma = M \, \partial u/\partial x$, so

$$\frac{\partial\sigma}{\partial x} = M \, \frac{\partial^2 u}{\partial x^2}$$

and

$$\frac{\partial^2 u}{\partial x^2} = \frac{\rho}{M} \, \frac{\partial^2 u}{\partial t^2} \qquad (2.6)$$

which is the standard form of the wave equation with the displacement wave velocity $v_s = (M/\rho)^{\frac{1}{2}}$. The displacement condition for static equilibrium is

$$\frac{\partial^2 u}{\partial x^2} = 0 \qquad (2.7)$$

and if a displacement function satisfies the equilibrium condition plus the boundary conditions of the problem, it is unique, because the minimum possible energy is associated with it, as is proved in elasticity texts.

Solutions of Eq. (2.6) have the general form $f(x \pm ct)$; that is, the displacement propagates with a velocity c, and substitution of the general form into (2.6) shows that $c^2 = M/\rho$.

A one-dimensional wave can be described most compactly by means of the complex exponential notation

$$u = U e^{i(kx - \omega t)} \qquad (2.8)$$

where k is the wave number $2\pi/\lambda$ and ω is the angular frequency $2\pi f$ (λ and f are the wavelength and linear frequency, respectively). Such a wave can be used to demonstrate the effects of *boundary conditions* and to indicate the importance of the *acoustic impedance*.

Suppose that at some point such as $x = 0$ the properties of a bar suddenly change from modulus M and density ρ to M' and ρ', respectively. A wave that is incident on the boundary, as in Fig. 2.3, with amplitude I is partially

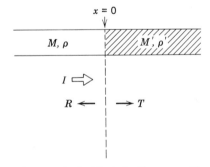

Fig. 2.3   An incident elastic wave splits into a transmitted part and a reflection when it strikes a sharp interface.

reflected with an amplitude R and partially transmitted with an amplitude T.

Since the event can occur at any time, t may be set equal to zero to eliminate the time factor. Also, two boundary conditions must be considered in order to evaluate the amplitudes R and T. One is that the displacement on the left must equal that on the right:

$$u = u' \qquad \text{at } x = 0$$

Therefore,

$$I + R = T \tag{2.9}$$

The other boundary condition is that the stresses on the two sides of the boundary must be equal so the forces are in equilibrium. This means that

$$\sigma = M \frac{\partial u}{\partial x} = + M' \frac{\partial u'}{\partial x} = +\sigma'$$

so that

$$Mk(I - R) = +k'M'T \tag{2.10}$$

Now a new quantity is defined that is called the acoustic impedance Z. It is equal to $\sqrt{M\rho} = c\rho = M/c = Mk/\omega$. Therefore Eq. (2.10) becomes

$$Z(I - R) = Z'T \tag{2.11}$$

and simultaneous solution of Eqs. (2.9) and (2.11) yields

$$R = \frac{Z - Z'}{Z + Z'} I$$
$$T = \frac{2Z}{Z + Z'} I \tag{2.12}$$

These coefficients refer to particle displacements or velocities, but the stresses in the waves are usually of more interest. The stress reflection and transmission coefficients are as follows:

$$R_\sigma = \frac{\sigma_R}{\sigma_I} = \frac{-R}{I} = \frac{Z' - Z}{Z' + Z} \tag{2.13}$$

$$T_\sigma = \frac{\sigma_T}{\sigma_I} = \frac{Z'}{Z} T = \frac{ZZ'}{Z' + Z} \tag{2.14}$$

Equations (2.13) and (2.14) indicate that if two different materials (that is, $M \neq M'$ and $\rho \neq \rho'$) with equal impedances exist on the left and right of the interface at $x = 0$, then $R = 0$, and $T = 1$, so an incident wave passes through without any reflection. In other words, elastic energy is transmitted with maximum efficiency. On the other hand, if a hard, dense material such as tungsten carbide lies at the right and a soft, light material such as Lucite is on the left, then $Z' \gg Z$, so $R_\sigma \simeq 1$, and $T_\sigma \simeq 0$. Thus most of an incident wave from the left gets reflected at the interface without change of sign. However, if the wave passes from the carbide into the polymer, $R_\sigma \simeq -1$, and again $T_\sigma \simeq 0$. Thus a strong reflection occurs in which the sign changes. In both cases the energy transmission is poor because of the impedance mismatch.

At a unit area of wavefront, the *power* P being generated is the stress times the particle velocity, since this gives the rate at which work is done. Therefore,

$$P = \sigma v = M\epsilon v = -Mk\omega(\mathrm{Re}\, u)^2$$
$$= \frac{-M\omega^2 U^2}{c} \sin^2 (kx + \omega t) \tag{2.15}$$

The time average of this at a particular place (say $x = 0$) is called the *intensity I:*

$$I = \langle P \rangle = \frac{M\omega^2 U^2}{2c} \tag{2.16}$$

Thus the intensity depends on the stiffness of the material, the square of the amplitude, and the frequency. The intensity can also be written in terms of the stress amplitude σ_0:

$$I = \frac{c}{2M} \sigma_0^2 \tag{2.17}$$

This completes our review of one-dimensional elastic behavior.

Solids typically extend in all three dimensions so that stresses and strains in them refer not to individual points but to small volumes of material. A precise description of them requires the use of special mathematical objects known as *tensors* that were first introduced by Voigt (1910). He gave them their name because he was concerned with describing a state of "tension," or system of elastic stresses.

Now that we have reviewed some of the elementary ideas about elastic behavior, we shall turn to the general three-dimensional description.

2.2 THE STRESS TENSOR

Stress is defined as a force per unit area, so the *state of stress* at a particular place in a three-dimensional solid describes the forces that act on an arbitrarily oriented element of surface inside the solid. An elemental area of this kind is sketched in Fig. 2.4. The force $\Delta \mathbf{F}$ can arise in various ways: (1) as a

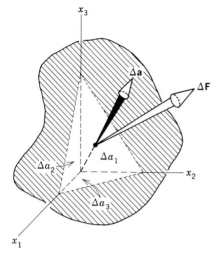

Fig. 2.4 Drawing of an elemental area
Δa in a solid that is acted upon by a force
ΔF (balanced by $-\Delta F$).

result of forces that act on the external surface (called *tractions*); (2) from local particle accelerations; (3) from an interaction of the material with electric, magnetic, or gravitational fields. Also, internal strains associated with dislocations or temperature differences will produce internal forces.

The force ΔF may be resolved into three orthogonal components ΔF_i ($i = 1,2,3$), and the area Δa has the components Δa_j ($j = 1,2,3$). Each of the three components of ΔF acts on an area component such as Δa_3. Thus there are three stresses σ_{i3} or forces per unit area acting on Δa_3:

$$\frac{\Delta F_1}{\Delta a_3} = \sigma_{13} \qquad \frac{\Delta F_2}{\Delta a_3} = \sigma_{23} \qquad \frac{\Delta F_3}{\Delta a_3} = \sigma_{33}$$

(Similar sets of stresses act on Δa_1 and Δa_2.) One of these (σ_{i3}) acts normal to Δa_3, and two act on it tangentially, as shown schematically in Fig. 2.5. In each case the first subscript indicates the direction of the force, and the second one gives the direction of the normal to the area.

In just the same way, three other stress components σ_{12}, σ_{22}, and σ_{32} can describe the forces per unit area that ΔF exerts on the surface element Δa_2 that lies normal to the x_2 axis. Also, the surface element Δa_1 has three stress components σ_{11}, σ_{22}, and σ_{31} acting on it. In all, nine stress components are needed to describe all the forces acting on the elemental cube (plus nine balancing ones).

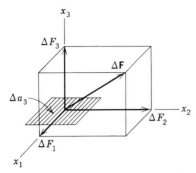

Fig. 2.5 Forces acting on an elementary surface area that lies normal to the x_3 axis. Note that $\Delta\mathbf{F}$ can be resolved into one normal component and two tangential ones.

The nine stress components defined above may be written:

$$\sigma_{ij} = \begin{bmatrix} \sigma_{11} & \sigma_{12} & \sigma_{13} \\ \sigma_{21} & \sigma_{22} & \sigma_{23} \\ \sigma_{31} & \sigma_{32} & \sigma_{33} \end{bmatrix} \qquad \begin{matrix} i = 1, 2, \text{ or } 3 \\ j = 1, 2, \text{ or } 3 \end{matrix} \qquad (2.18)$$

They form the components of a second-rank tensor, which means that, although the individual values of the components of this tensor will change if the choice of coordinate axes is changed, the tensor as a whole will not be affected. Proofs of this are available in the books of the reference list. Its plausibility is evident, since the σ_{ij} are simply coefficients that connect two vectors (force and area) and the independence of a vector from the particular set of coordinates that is used to describe it can be readily visualized.

If an arbitrary surface element (defined by a normal vector $\mathbf{n}$ whose components are n_i) is given, the stress tensor will determine the forces acting on it. These forces are three in number and may be written:

$$\sigma_{in} = \sum_{j} \sigma_{ij} n_j \qquad \begin{matrix} i = 1, 2, \text{ or } 3 \\ j = 1, 2, \text{ or } 3 \end{matrix} \qquad (2.19)$$

In extended form the force components (assuming unit area) are

$$\sigma_{1n} = \sigma_{11} n_1 + \sigma_{12} n_2 + \sigma_{13} n_3$$
$$\sigma_{2n} = \sigma_{21} n_1 + \sigma_{22} n_2 + \sigma_{23} n_3$$
$$\sigma_{3n} = \sigma_{31} n_1 + \sigma_{32} n_2 + \sigma_{33} n_3$$

Because the tangential surface forces create force couples which must be counteracted by equal but oppositely directed couples in order to avoid rotation of an element, the stress tensor is symmetric. A study of Fig. 2.6 should make this clear. Here all the *force* pairs are in balance, but the *couples* caused by the σ_{12} and σ_{21} pairs are not balanced unless they are equal. There-

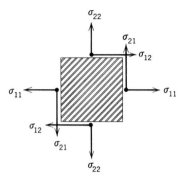

Fig. 2.6   Cross section of an internal volume element, showing the forces that act on it.

fore, for static equilibrium,

$$\sigma_{12} = \sigma_{21}$$

and a similar argument leads to the general relation

$$\sigma_{ij} = \sigma_{ji} \qquad i,j = 1, 2, \text{ or } 3$$

The result of this symmetry is that the stress tensor takes the form

$$\begin{bmatrix} \sigma_{11} & \sigma_{12} & \sigma_{13} \\ \sigma_{12} & \sigma_{22} & \sigma_{23} \\ \sigma_{13} & \sigma_{23} & \sigma_{33} \end{bmatrix}$$

where the diagonal terms describe tensions and compressions and the off-diagonal terms describe shears. Pressure or hydrostatic tension is given by the average of the diagonal terms

$$P = \pm\tfrac{1}{3}(\sigma_{11} + \sigma_{22} + \sigma_{33})$$

and if they are all equal, the state of stress is one of pure pressure or hydrostatic tension.

In general, a state of stress imposes both pressure (dilatational) and shear forces. Since it is easier to visualize these separately than it is to visualize the complete tensor, it is commonly resolved into a *spherical* (σ_{ij}^S) and a *deviator* (σ_{ij}^D) part. The spherical part is

$$
\sigma_{ij}^S = \begin{bmatrix} P & 0 & 0 \\ 0 & P & 0 \\ 0 & 0 & P \end{bmatrix}
\tag{2.20}
$$

whereas the deviator (or shear) part is

$$
\sigma_{ij}^D = \begin{bmatrix} \dfrac{2\sigma_{11} - (\sigma_{22} + \sigma_{33})}{3} & \sigma_{12} & \sigma_{13} \\[2ex] \sigma_{12} & \dfrac{2\sigma_{22} - (\sigma_{11} + \sigma_{33})}{3} & \sigma_{23} \\[2ex] \sigma_{13} & \sigma_{23} & \dfrac{2\sigma_{33} - (\sigma_{11} + \sigma_{22})}{3} \end{bmatrix}
\tag{2.21}
$$

Through the choice of a suitable set of reference axes, the stress tensor can always be reduced to the form

$$
\sigma_{ij} = \begin{bmatrix} \sigma_1 & 0 & 0 \\ 0 & \sigma_2 & 0 \\ 0 & 0 & \sigma_3 \end{bmatrix} \qquad \sigma_1 > \sigma_2 > \sigma_3
$$

where the remaining components are called the *principal stresses* and these act in the three *principal directions*.

This completes a review of how the state of stress in a material is described. Next, a means for describing the state of strain is needed.

2.3 THE STRAIN TENSOR

When a solid is deformed, points in it are displaced by various amounts that are defined by displacement vectors **u**. If a given point that is defined initially by a vector $\mathbf{r}_0$ (as in Fig. 2.7) is displaced to a position defined by a vector **r**, then the displacement is given by

$$
\mathbf{u} = \mathbf{r} - \mathbf{r}_0
\tag{2.22}
$$

and the vectors **u** form a vector displacement field.

The vector displacements of nearby points cannot be completely arbitrary, as this might cause holes to open up or more than one particle to occupy the same place. Therefore, there are constraints known as *compatibility relations* that are described in standard texts.

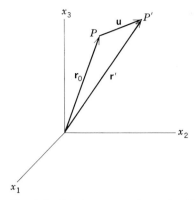

Fig. 2.7 Displacement of a point P to a new position P'.

The displacements within a small volume element are given by

$$u_1 = u_1^0 + \frac{\partial u_1}{\partial x_1} x_1 + \frac{\partial u_1}{\partial x_2} x_2 + \frac{\partial u_1}{\partial x_3} x_3$$

$$u_2 = u_2^0 + \frac{\partial u_2}{\partial x_1} x_1 + \frac{\partial u_2}{\partial x_2} x_2 + \frac{\partial u_2}{\partial x_3} x_3 \qquad (2.23)$$

$$u_3 = u_3^0 + \frac{\partial u_3}{\partial x_1} x_1 + \frac{\partial u_3}{\partial x_3} x_2 + \frac{\partial u_3}{\partial x_3} x_3$$

where the $\partial u_i/\partial x_j$ are components of a *deformation tensor* D_{ij}. If the deformation is *homogeneous*, then it is independent of the origin of coordinates u_i^0 and this vector can be neglected. For small deformations, homogeneity is a reasonable assumption, but if they become finite, then changes of u_i^0 must be considered. If the local situation of the small element is considered, the following relation may be written:

$$\Delta u_i = \sum_j D_{ij} \Delta x_j \qquad (2.24)$$

It may be seen that the deformation tensor connects two vectors: one is the initial shape of the element; the other is the displacement of the shape. For a constrained element, all the components of D_{ij} lead to internal distortions and hence to changes of the internal energy. However, some of the components describe a uniform rotation of the entire element, and if the element is unconstrained, the atomic particles within it retain their positions relative to one another, so no internal energy change occurs. Because of this, and in spite of the fact that it leads to confusion, it is *customary* to split the deformation tensor into two parts, a *rotation tensor* ω_{ij} and a *strain tensor* ϵ_{ij}.[†]

The rotational displacements are given by

$$\Delta u_i^R = \sum_j \omega_{ij} \, \Delta x_j \tag{2.25}$$

and the strain displacements by

$$\Delta u_i^S = \sum_j \epsilon_{ij} \, \Delta x_j \tag{2.26}$$

where $\omega_{ij} = \dfrac{1}{2}\left(\dfrac{\partial u_j}{\partial x_i} - \dfrac{\partial u_i}{\partial x_j}\right) \qquad i,j = 1, 2, \text{ or } 3$

$$= \begin{bmatrix} 0 & -\omega_{21} & +\omega_{31} \\ +\omega_{12} & 0 & -\omega_{32} \\ -\omega_{13} & +\omega_{23} & 0 \end{bmatrix} \tag{2.27}$$

$\epsilon_{ij} = \dfrac{1}{2}\left(\dfrac{\partial u_j}{\partial x_i} + \dfrac{\partial u_i}{\partial x_j}\right) \qquad i,j = 1, 2, \text{ or } 3$

$$= \begin{bmatrix} \epsilon_{11} & \epsilon_{12} & \epsilon_{13} \\ \epsilon_{21} & \epsilon_{22} & \epsilon_{23} \\ \epsilon_{31} & \epsilon_{32} & \epsilon_{33} \end{bmatrix} \tag{2.28}$$

The strain tensor can be further split into two parts

$$\epsilon_{ij} = \begin{bmatrix} \epsilon_{11} & 0 & 0 \\ 0 & \epsilon_{22} & 0 \\ 0 & 0 & \epsilon_{33} \end{bmatrix} + \begin{bmatrix} 0 & \epsilon_{12} & \epsilon_{13} \\ \epsilon_{21} & 0 & \epsilon_{23} \\ \epsilon_{31} & \epsilon_{32} & 0 \end{bmatrix} \tag{2.29}$$

where the first term describes uniform dilatations or compressions, and the second describes shears.

Thus there are three general classes of changes in the configuration of a solid when it becomes deformed: (1) rotation, (2) stretching (or compressing), and (3) shearing. These are illustrated in Fig. 2.8, which also indicates at (*a*) that stretching is described by homogeneous terms of the form $\Delta u_i/\Delta x_j$ that give displacement per unit length parallel to the length; at (*b*) that shears are described by the addition of mixed terms of the form $\Delta u_j/\Delta x_i + \Delta u_i/\Delta x_j$ that define the shear angle; and at (*c*) that rotations are described by the subtraction of mixed terms of the form $\Delta u_j/\Delta x_i - \Delta u_i/\Delta x_j$ that define the rotation angle.

Shears and dilatations cause changes of distances between atomic particles and of the angles between three or more of them. In contrast, rotations do not cause such changes. Because of this behavior, it is commonly assumed

[†] Note that a bar is placed under ϵ to distinguish components of the strain tensor from "engineering strain components" $|\underline{\epsilon}| = \frac{1}{2}|\epsilon|$.

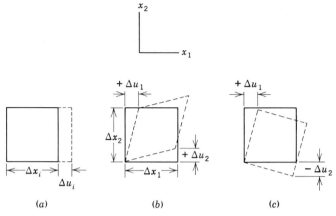

Fig. 2.8   Illustration of the three kinds of configuration changes
that relative displacements can produce in a solid.

that shears and dilatations cause internal energy changes, whereas rotations
do not. However, if a solid contains long-range internal force fields (such as
those connected with spontaneous magnetization or polarization), then
material that becomes rotated at some point P_1 may interact with nonrotated
material at another point P_2. Because of this possibility, the traditional
assumption is not always valid, but it has determined the standard formalism
that is used to describe the local strain.

The strain tensor has only six independent terms, since $\epsilon_{ij} = \epsilon_{ji}$ by definition. Hence,

$$\epsilon_{ij} = \begin{bmatrix} \epsilon_{11} & \epsilon_{12} & \epsilon_{13} \\ \epsilon_{21} & \epsilon_{22} & \epsilon_{23} \\ \epsilon_{31} & \epsilon_{32} & \epsilon_{33} \end{bmatrix} = \begin{bmatrix} \epsilon_{11} & \epsilon_{12} & \epsilon_{13} \\ \epsilon_{12} & \epsilon_{22} & \epsilon_{23} \\ \epsilon_{13} & \epsilon_{23} & \epsilon_{33} \end{bmatrix} \tag{2.30}$$

For a *homogeneous* state of strain, the ϵ_{ij} are constants independent of position, so the components of the displacement vectors can be described as
fractions of the position vectors. For example,

$$u_1 = \epsilon_{12}x_1 + \epsilon_{12}x_2 + \epsilon_{13}x_3$$

or, more generally,

$$u_i = \sum_j \epsilon_{ij}x_j \tag{2.31}$$

However, the more common case of an *inhomogeneous* state of strain will
not allow this simplification, because only the local situation can be described

in terms of the ϵ_{ij}. Strain gradients will be present which cause material at one position to be rotated (as well as translated) relative to material at another. Therefore displacement changes are related to position changes by *both* strains and rotations:

$$\Delta u_i = \sum_j (\epsilon_{ij} - \omega_{ij}) \Delta x_j \qquad (2.32)$$

2.4 THE ELASTICITY TENSOR

The elastic rigidity of a solid is the property that clearly distinguishes it from other forms of matter, and it is not surprising that the formal theory is highly developed. The plan of this section is first to introduce the elastic formalism, and then to describe how the elastic constants depend on chemical constitution and to briefly outline the atomic theory of elastic stiffness.

Compared with an isotropic elastic continuum, a crystal is a complex mechanical network of attractive and repulsive forces. In crystals with layered structures, such as mica, gypsum, and graphite, the elastic properties can be exceedingly anisotropic. On the other hand, for some metals, such as aluminum and tungsten, the elastic properties are nearly, or exactly, isotropic, depending on the measuring temperature.

The simplest kind of statement of Hooke's law is that for reversible strains, stress and strain are linearly proportional to one another. For wires in uniaxial tension or torsion, this statement is adequate as it stands, but for three-dimensional stress systems, and especially in the anisotropic case, considerable elaboration is needed.

Since a state of stress can always be referred to a set of three principal axes that are orthogonal, it is convenient to define the stresses in a generalized Hooke's law relative to orthogonal cartesian coordinates. The crystal structure is defined, however, relative to three crystal axes that are not necessarily orthogonal. Therefore, it is necessary to establish a conventional set of relationships between the crystal axes and the cartesian coordinates. This will be done after Hooke's law is generalized.

The stress and strain tensors are separately independent of the material in which they exist. According to Hooke, however, their components are proportional to each other within a given material, and the proportionality constants vary from one material to another (and often from one direction to another within one material).

Since stress and strain are tensors, the set of proportionality coefficients that connects them is also a tensor. It is of increased rank because 81 coefficients are needed to connect nine components each of stress and of strain. The generalized tensor form of Hooke's law is

$$\sigma_{ij} = \sum_{kl} C_{ijkl}\varepsilon_{kl} \qquad i,j,k,l = 1, 2, \text{ or } 3 \tag{2.33}$$

and the coefficients C_{ijkl} form the *elastic stiffness tensor*. As the stiffness of a solid increases, so do the magnitudes of the C_{ijkl} coefficients.

The proportionality of stress and strain can also be expressed as

$$\varepsilon_{ij} = \sum_{kl} S_{ijkl}\sigma_{kl} \tag{2.34}$$

and the S_{ijkl} form the *elastic compliance tensor*. This form of Hooke's law often has more practical utility than the former one, because it is easier to apply a single stress component and measure all the resulting strains than it is to apply a strain and measure the resulting stresses. If the compliance tensor is known, the stiffness tensor can be calculated, because the two tensors are inversely related to one another. This may be written in terms of the Kronecker delta function as follows:

$$C_{ijkl}S_{ijmn} = \tfrac{1}{2}(\delta_{km}\delta_{ln} + \delta_{kn}\delta_{lm}) = \delta_{kn}\delta_{lm}$$
$$\delta_{km} = \begin{array}{ll} 0 & \text{if } k \neq m \\ 1 & \text{if } k = m \end{array} \tag{2.35}$$

If the elasticity tensor of a material is known, it is possible to write a general expression for the strain energy of an elastically strained solid. By analogy with the one-dimensional case discussed previously, the strain energy density is equal to the reversible work that is done in straining a material and is given at each point by

$$W = \frac{1}{2} \sum_{ijkl} C_{ijkl}\varepsilon_{ij}\varepsilon_{kl} \tag{2.36}$$

Integration of this over the entire volume gives the total stored energy

$$W = \frac{1}{2} \int \left(\sum_{ijkl} C_{ijkl}\varepsilon_{ij}\varepsilon_{kl} \right) dV \tag{2.37}$$

For any set of boundary tractions or internal forces, there is a field of displacements that minimizes W. Therefore, the correct displacement field is a unique solution for the given set of boundary conditions.

The search for the solution of an elasticity problem is greatly aided because it can be shown that a field of displacements that is consistent with the boundary tractions and internal forces plus the equilibrium conditions automatically minimizes the strain energy integral. The generalized equilibrium equations (when no internal forces act) are

$$\sum_j \frac{\partial \sigma_{ij}}{\partial x_j} = 0 \qquad i,j = 1, 2, \text{ or } 3 \tag{2.38}$$

2.5 EFFECT OF SYMMETRY ON THE ELASTICITY TENSOR

The elastic stiffnesses and compliances relate two tensors (σ_{ij} and ϵ_{ij}) which each have 9 terms. Hence C_{ijkl} and S_{ijkl} each have 81 terms altogether. However, both σ_{ij} and ϵ_{ij} are symmetric so the elasticity tensors are also. Now a 9×9 matrix has 9 diagonal terms, leaving 72 off-diagonal terms; symmetry makes these equal in pairs, leaving 36. Thus there are 45 different terms to begin with.

In addition, equilibrium requirements demand that $\sigma_{ij} = \sigma_{ji}$ and ϵ_{ij} and ϵ_{ji}, reducing each tensor to 6 independent terms and eliminating 3 diagonal terms as well as 21 off-diagonal ones. The net remainder is 21 independent terms.

In addition to the general symmetries discussed above, elastic substances may possess internal-symmetry elements, and these can cause further simplification of the elasticity tensors. For example, a crystal that has cubic symmetry must look the same (elastically) when it is viewed along the x_1 axis as when it is viewed along x_2 or x_3. Therefore, it cannot have more than three independent elastic constants (one to describe stretching and two for shear, for example).

Before these internal-symmetry effects can be discussed clearly, we need to describe the nature of the C_{ijkl} as they apply to crystals. As mentioned above, a complication arises because the C_{ijkl} (or S_{ijkl}) are referred to orthogonal axes (x_1, x_2, x_3), whereas the crystal axes (a, b, c) are nonorthogonal in general. Therefore, a convention is required to link the a, b, c to the x_1, x_2, x_3. Standard conventions have been specified by the Institute of Electronic and Electrical Engineers, and we shall follow their standards. These are given pictorially in Fig. 2.9.

For some of the elastic constants, the nature of what happens when a stress or strain is applied is apparent, but for others, this may not be clear. This is especially true for low-symmetry crystals. Therefore, the strains that are described by various constants when certain stresses are applied have

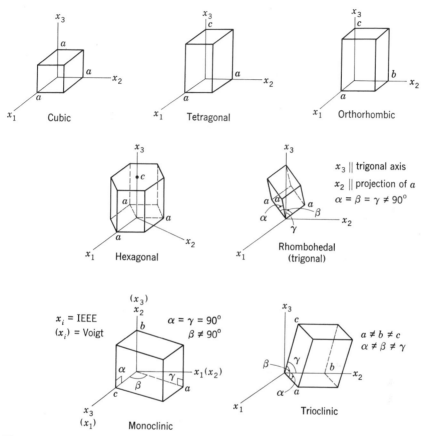

Fig. 2.9 Orientation conventions for crystals relative to three orthogonal axes (*IEEE Standards*).

been outlined in Fig. 2.10. Here the most familiar modes are labeled: L for extension caused by simple tension; P for lateral contraction caused by tension; and S for shear induced by simple shear. In low-symmetry crystals, it may be seen that a stress of one kind may induce a strain of another kind. For example, in the LS mode, an applied tension induces a shear strain.

The effect of the symmetries of crystals is to reduce the number of constants that is needed to describe the elastic behavior. In order to demonstrate this effect, a specific stiffness constant will be considered in the cubic system. This is C_{1123}, and its meaning is illustrated by Fig. 2.11, where it is assumed to be greater than zero. Then a strain $+\varepsilon_{23}$ causes a stress $+\sigma_{11}$. But suppose that the crystal is rotated about the x_3 axis so that $+x_2$ becomes

	σ_{11}	σ_{22}	σ_{33}	σ_{23}	σ_{31}	σ_{12}
ϵ_{11}	L	P	P	SP	SL	SP
ϵ_{22}	P	L	P	SP	SP	SL
ϵ_{33}	P	P	L	SL	SP	SP
ϵ_{23}	LS'	LS'	LS	S	SS'	SS'
ϵ_{31}	LS	LS'	LS'	SS'	S	SS'
ϵ_{12}	LS'	LS	LS'	SS'	SS'	S

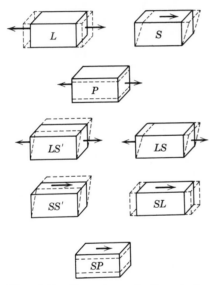

Fig. 2.10 Characters of the strains associated with the various elastic compliances.

$-x_2$. Then $+\epsilon_{23}$ becomes $-\epsilon_{23}$, although it does not change physically, because the x_1x_2 plane has mirror symmetry. Since $\sigma_{11} = C_{1123}\epsilon_{23}$, this means that σ_{11} is negative (compressive). But a change of coordinates cannot change the sense of a physical quantity. Therefore, it must be concluded that $C_{1123} = 0$ in the cubic system.

Furthermore, a cubic crystal is symmetric about all three planes: (x_1x_2), (x_2x_3), and (x_3x_1). Therefore, C_{ijkl} must be invariant for changes of sign in x_3 or x_1 or x_2. Thus all terms of the form q_{iiij} must equal zero. Only terms of the form C_{iijj} or C_{ijij} or C_{iiii} are nonzero.

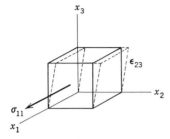

Fig. 2.11   Illustration of the stress σ_{11} that would result from a strain ϵ_{23} if the stiffness constant C_{1123} were finite for a cubic crystal.

Not all the terms of the latter form above are different, because symmetry creates certain equalities among them. A cubic crystal looks the same in the three directions x_1, x_2, x_3; that is, its behavior is independent of the transformations

$$x_1 \rightarrow x_2 \qquad x_1 \rightarrow x_3$$
$$x_2 \rightarrow x_3 \qquad x_2 \rightarrow x_1$$
$$x_3 \rightarrow x_1 \qquad x_3 \rightarrow x_2$$

Hence,

$$C_{1111} = C_{2222} = C_{3333}$$

and

$$C_{1122} = C_{3322} = C_{2211} = C_{1133} = C_{3311} = C_{2233}$$

as well as

$$C_{1212} = C_{2323} = C_{1313} = C_{2121} = C_{3232} = C_{3131}$$

Therefore, the matrix of coefficients contains only three independent constants and has the following appearance:

$$
\begin{array}{ccc|ccc}
C_{1111} & C_{1122} & C_{1122} & 0 & 0 & 0 \\
C_{1122} & C_{1111} & C_{1122} & 0 & 0 & 0 \\
C_{1122} & C_{1122} & C_{1111} & 0 & 0 & 0 \\
\hline
0 & 0 & 0 & C_{1212} & 0 & 0 \\
0 & 0 & 0 & 0 & C_{1212} & 0 \\
0 & 0 & 0 & 0 & 0 & C_{1212}
\end{array}
$$

For an *isotropic* material the constants must also be independent of *any* rotation of the coordinate system. This means that

$$C_{1111} = C_{1122} + 2C_{1212}$$

which reduces the number of independent constants to two.

Consider the constants needed for a case more complex than cubic. The result for the case of the orthorhombic system can be readily visualized, since the crystal axes are orthogonal. Thus, three constants are needed to describe dilatations along the three crystal axes; these are C_{1111}, C_{2222}, and C_{3333}. Then, three shear constants are needed to describe shears in the directions of the three axes; these are C_{2323}, C_{3131}, and C_{1212}. Finally, three constants are needed to describe transverse strains; these are C_{1122}, C_{1133}, and C_{2233}. The total then is nine.

The numbers of independent constants for the various crystal systems are as follows:

Isotropic: 2
Cubic: 3
Hexagonal: 5
Tetragonal: 6
Rhombohedral: 6
Orthorhombic: 9
Monoclinic: 13
Triclinic: 21

It is often convenient to assume that crystals are nearly isotropic, because this simplifies a discussion of the properties. In order to estimate whether this is a good or bad assumption, it is desirable to have a quantitative means for specifying the degree of anisotropy. For crystals that have polar axes (tetragonal, rhombohedral, hexagonal), deviations from isotropy are obvious from an inspection of the elastic constants parallel and perpendicular to the polar axis. For cubic crystals, the three axes are equivalent, so the deviations from isotropy are more subtle but can be given a simple analytic definition.

In order to define what is meant by the anisotropy of a cubic crystal, it is desirable to first find an expression for the Young's modulus in an arbitrary direction. This modulus describes the stress associated with a single strain component

$$\sigma_{ii} = S_{iiii}\epsilon_{ii}$$

so if S'_{1111} is the tensile compliance in an arbitrarily rotated coordinate system x'_1, x'_2, x'_3, then the Young's modulus Y_{hkl} in an arbitrary direction with the Miller indices hkl is

$$Y_{hkl} = \frac{1}{S'_{1111}} \tag{2.39}$$

and the problem becomes that of finding S'_{1111}.

Let α_{ij} be the direction cosines between the axes of the x_i coordinate system and the x'_i system. Then the direction cosines form the following matrix connecting the two coordinate systems:

	x_1	x_2	x_3
x'_1	α_{11}	α_{12}	α_{13}
x'_2	α_{21}	α_{22}	α_{23}
x'_3	α_{31}	α_{32}	α_{33}

Thus, a given vector with the components b_j in the x_i system has the following components in the x'_i system:

$$b'_i = \sum_j \alpha_{ij} b_j \tag{2.40}$$

In textbooks (Nye, 1960) it is shown that a second-rank tensor T_{kl} in the x_i system has the following components in the x'_i system:

$$T'_{ij} = \sum_{kl} \alpha_{ik} \alpha_{jl} T_{kl} \tag{2.41}$$

Furthermore, a fourth-rank tensor may be transformed from one system to another by means of the following expression for the components:

$$S'_{ijkl} = \sum_{mnop} \alpha_{im} \alpha_{jn} \alpha_{ko} \alpha_{lp} S_{mnop} \tag{2.42}$$

or, in the case of interest here:

$$\begin{aligned}
S'_{1111} &= \sum_{mnop} \alpha_{lm} \alpha_{ln} \alpha_{lo} \alpha_{lp} S_{mnop} \\
&= \alpha_{11}^4 S_{1111} + \alpha_{12}^4 S_{2222} + \alpha_{13}^4 S_{3333} + \alpha_{11}^2 \alpha_{12}^2 S_{1122} \\
&\quad + \alpha_{11}^2 \alpha_{13}^2 S_{1133} + \alpha_{12}^2 \alpha_{11}^2 S_{2211} + \alpha_{12}^2 \alpha_{13}^2 S_{2233} + \alpha_{13}^2 \alpha_{11}^2 S_{3311} \\
&\quad + \alpha_{13}^2 \alpha_{12}^2 S_{3322} + 4\alpha_{12}^2 \alpha_{13}^2 S_{2323} + 4\alpha_{11}^2 \alpha_{13}^2 S_{1313} + 4\alpha_{11} \alpha_{12}^2 S_{1212}
\end{aligned}$$

If the standard relations for the sums of functions of the direction cosines are applied,

$$\alpha_{11}^2 + \alpha_{12}^2 + \alpha_{13}^2 = 1 \qquad \alpha_{12}\alpha_{13} + \alpha_{22}\alpha_{23} + \alpha_{32}\alpha_{33} = 0$$
$$\alpha_{21}^2 + \alpha_{22}^2 + \alpha_{23}^2 = 1 \qquad \alpha_{13}\alpha_{11} + \alpha_{23}\alpha_{21} + \alpha_{33}\alpha_{31} = 0$$
$$\alpha_{31}^2 + \alpha_{32}^2 + \alpha_{33}^2 = 1 \qquad \alpha_{11}\alpha_{12} + \alpha_{21}\alpha_{22} + \alpha_{31}\alpha_{32} = 0$$

The result may be written

$$S'_{1111} = S_{1111} - (2S_{1111} - 2S_{1122} - 4S_{2323})(\alpha_{11}^2\alpha_{12}^2 + \alpha_{11}^2\alpha_{13}^2 + \alpha_{12}^2\alpha_{13}^2) \quad (2.43)$$

and for the three most important directions in a cubic crystal, the direction cosines are

$$\langle 100 \rangle \alpha_{ij} = 1, 0, 0$$
$$\langle 110 \rangle \alpha_{ij} = \frac{1}{\sqrt{2}}, \frac{1}{\sqrt{2}}, 0$$
$$\langle 111 \rangle \alpha_{ij} = \frac{1}{\sqrt{3}}, \frac{1}{\sqrt{3}}, \frac{1}{\sqrt{3}}$$

so the Young's moduli are

$$\frac{1}{Y_{100}} = S_{1111}$$

$$\frac{1}{Y_{110}} = S_{1111} - \tfrac{1}{2}\left[(S_{1111} - S_{1122}) - 2S_{2323}\right] \qquad (2.44)$$

$$\frac{1}{Y_{111}} = S_{1111} - \tfrac{2}{3}\left[(S_{1111} - S_{1122}) - 2S_{2323}\right]$$

The condition for isotropy is $Y_{100} = Y_{110} = Y_{111}$, which requires that the term in brackets be zero; that is, $S_{1111} - S_{1122} = 2S_{2323}$, and the anistropy factor A is a measure of deviations from this equality:

$$A = \frac{S_{1111} - S_{1122}}{2S_{2323}} \qquad (2.45)$$

This also has a simple physical interpretation, because the numerator measures the compliance for a shear directed along a cube-face diagonal, whereas the denominator refers to a shear directed along a cube edge.

Examples of anistropy factors for some metals, salts, and covalent crystals are given in Table 2.1.

During the early development of the atomic theory of the elastic stiffnesses of materials, it was not known whether the forces that act between

TABLE 2.1 EXAMPLES OF ELASTIC ANISOTROPY FACTORS FOR CUBIC CRYSTALS

CRYSTAL	ANISOTROPY FACTOR $= A$
Al	1.23
Fe	2.40
Au	3.90
W	1.00
Na	7.50
NaCl	0.70
KBr	0.35
KI	0.36
Ge	1.66
Si	1.56

atoms are always directed along lines that pass through their centers. A controversy arose, and the issue was analyzed by the famous mathematician Cauchy, who showed that if the forces act only along central lines, certain additional conditions on the elastic constants must hold. These are known as the Cauchy relations, and the one for cubic crystals is

$$C_{1122} = C_{2323} \tag{2.46}$$

A simplified proof of this may be found in the book by Feynman, Leighton, and Sands (1964, pp. 10–13, 39).

In order to test the Cauchy relations, Voigt (1887) made careful measurements of the elastic constants of several salts and found that they are sometimes, but often not, obeyed. This is demonstrated in Table 2.2 and leads to the conclusion that the forces between atoms do not act only along lines connecting their centers.

2.6 THE CHEMISTRY OF THE ELASTIC CONSTANTS

Elastic stiffness measures the rate at which the binding energy of a crystal increases when it is stretched or sheared at $0°K$. In general, cohesion results from electrostatic attractions between atomic particles. In salts the particles are ions that carry different charges; in metals the positive ion cores are

**TABLE 2.2 COMPARISON OF ELASTIC STIFFNESS CONSTANTS
THAT WOULD BE EQUAL IF ONLY CENTRAL FORCES ACTED
BETWEEN ATOMS** (*Cauchy theory*)

CRYSTAL	C_{1122} (10^{11} dyn/cm²)	C_{2323} (10^{11} dyn/cm²)
NaCl	1.24	1.26
NaBr	1.31	1.33
KCl	0.62	0.62
KBr	0.54	0.51
KI	0.58	0.62
LiF	4.20	6.20
MgO	8.70	14.8
Cu	12.3	7.53
Ag	8.97	4.36
Au	15.7	4.2
Al	6.13	2.85
Na	0.46	0.59
W	19.8	15.1
C	12.50	57.6
Si	6.40	7.96
Ge	4.84	6.72

attracted to the negative electron gas; and in a covalent crystal, electron
pair bonds attract their associated ions.

According to Coulomb's law, oppositely charged particles are attracted
by forces proportional to e^2/r^2, where e is the amount of charge and r is the
interparticle spacing.

Since crystals are close-packed, the area across which the cohesive force
acts is proportional to r^2. Therefore, the cohesive stress (elastic stiffness)
should be proportional to e^2/r^4. Indeed, for rock salt (NaCl), the stiffness
roughly equals this quantity:

$$\frac{e^2}{r^4} (\text{NaCl}) = \frac{(4.8 \times 10^{-10})^2}{(2.8 \times 10^{-8})^4} \simeq 3.8 \times 10^{11} \text{ dyn/cm}^2$$

compared with the experimental value of C_{1111}, which is 4.9×10^{11} dyn/cm².
This result indicates that the most important atomic parameters for high
elastic stiffness are large numbers of charges participating in the binding and
short interparticle distances.

It is not the plan here to discuss the detailed physical theory of elastic
stiffness. Such a discussion involves so many assumptions that it becomes
doubtful whether it generates insight commensurate with effort. Instead, as
the title of this section implies, an attempt will be made to identify the

atomic properties that influence elastic stiffness, so that if one species is substituted for another, the result can be anticipated semiquantitatively.

The first crystals for which an atomic theory of cohesion was developed are the salts. Even before their structures were definitely known, Madelung (1909) had proposed that salt crystals consist of discrete ions with alternating signs, organized in a periodic pattern as proposed by Barlow (1898). The evidence that discrete ions persisted in the solid state was that infrared radiation of the appropriate frequency to excite ionic oscillators is strongly absorbed by these crystals.

Born (1923) developed the mathematics of Madelung's model and made detailed comparisons with experimental measurements. These comparisons were remarkably successful, so the theory became the basis of the theory of solids. It is assumed in this model that the ions are spherically symmetric and that they interact along lines that connect their centers. The most important interaction between them is a net electrostatic attraction which tends to contract the assembly of particles. However, at close distances their cores begin to interact, and because of the Pauli exclusion principle, this causes a rapid increase in the kinetic energies of the core electrons. Since the electrons have distributions that decrease exponentially with distance from their centers, the energies tend to increase exponentially with decreasing distance. The two effects (attraction and repulsion) may be described by means of the Born-Mayer interionic potential function

$$\Phi_{mn} = \pm \frac{e^2}{r_{mn}} + A e^{-r_{mn}/\rho}$$

where the first term comes from Coulomb's law and its sign depends on whether two ions have equal or opposite signs. In the second term, A is a constant and ρ is a relaxation radius which measures the size of the "hard spherical core."

The potential energy of a piece of crystal is

$$U = \frac{1}{2} \sum_m \sum_n \Phi_{mn} \qquad m \neq n$$

where the factor of $1/2$ compensates for the fact that each ion appears twice in the summation of the pairs. The density of energy in a crystal of volume V is $E = U/V$ and the elastic stiffnesses are then given by

$$C_{ijkl} = \left(\frac{\partial^2 E}{\partial \varepsilon_{ij} \, \partial \varepsilon_{kl}} \right)_{r=r_0} \tag{2.47}$$

where r_0 is the equilibrium spacing.

The constant A may be eliminated by using the condition that $\Phi(r)$ is a minimum at $r = r_0$, so that

$$\left(\frac{\partial \Phi}{\partial r}\right)_{r=r_0} = 0 = \mp \frac{e^2}{r_0^2} - \frac{A}{\rho} e^{-r_0/\rho}$$

and the expression for $\Phi(r_0)$ becomes

$$\Phi(r_0) = \pm \frac{e^2}{r_0}\left(1 - \frac{\rho}{r_0}\right) \tag{2.48}$$

Values for ρ can be determined from experimental values of the bulk modulus B which describes the pressure P needed to cause a given volume change

$$\frac{\Delta V}{V} = -\frac{P}{B} \tag{2.49}$$

and whose definition in terms of internal energy changes is

$$B = V_0 \left(\frac{\partial^2 U}{\partial V^2}\right)_{V=V_0} \tag{2.50}$$

It is found that ρ/r_0 lies in the range 0.09–0.12, so the largest contribution to Φ comes from the electrostatic term.

A treatment of the elastic constants by Krishnan and Roy (1952) isolates two important lattice sums. One is a sum over Θ_{mn}, which is known as Madelung's sum and is needed to obtain U. The other is a sum over $\partial^2 \Phi_{mn}/\partial r_{mn}^2$, which is part of $\partial^2 E/\partial \varepsilon_{ij}\,\partial \varepsilon_{kl}$ and will be called χ. These sums have the forms

$$\alpha = -\Sigma(-1)^s(x^2 + y^2 + z^2)^{-\frac{1}{2}}$$
$$\chi = -\Sigma(-1)^s(x^4 + y^4 + z^4)(x^2 + y^2 + z^2)^{-\frac{5}{2}}$$

where $x, y, z \neq 0$; $x, y, z = 1, 2, 3, 4 \ldots$; $s = x + y + z$.

TABLE 2.3 COMPARISON OF SOME MEASURED AND CALCULATED ELASTIC STIFFNESSES FOR SALT CRYSTALS (Units are 10^{11} dyn/cm²)

CRYSTAL	C_{1111}		$C_{1122} = C_{2323}$	
	MEAS.	CALC.	MEAS.	CALC.
NaCl	4.87	5.00	1.26	1.30
NaBr	3.87	4.30	0.97	1.00
KCl	3.98	3.90	0.63	0.80
KBr	3.46	3.50	0.51	0.70

For the rock salt structure they have the values $\alpha = 1.75$ and $\chi = 3.14$ and lead to the following values for the elastic constants:

$$C_{1122} = C_{2323} = 0.348 \frac{e^2}{r_0^4}$$

$$C_{1111} = [3.5(\delta + 1) - 18.8] \frac{e^2}{12r_0^4} \tag{2.51}$$

with $\delta = r_0/\rho$. Table 2.3 shows that values calculated on this basis compare very well with experimental measurements.

Equations (2.51) indicate that the shear stiffness depends only on the parameter e^2/r_0^4. Therefore, a plot of the logarithm of C_{2323} versus the logarithm of r_0 should yield a linear correlation with slope $= -4$, and Fig. 2.12 shows that this is indeed the case. All the data cluster about a trend

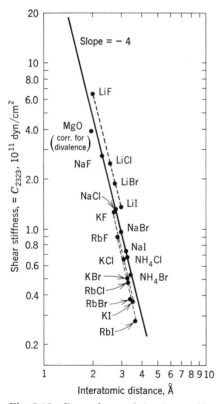

Fig. 2.12 Dependence of the shear stiffnesses of ionic crystals on interionic distances.

line that has the expected slope and whose position is given by the Krishnan-Roy theory. This has also been emphasized by Haussuhl (1963). However, factors other than the ones considered in the theory make small contributions which systematically increase the stiffnesses of the lithium salts and decrease those of the potassium and rubidium salts.

It may also be noted that if the experimental value of the shear stiffness of MgO is divided by 4 to correct for the divalence of its ions (the others in Fig. 2.12 being monovalent), then it falls close to the trend line of the figure.

The alkali halides are combinations of the most electropositive elements and the most electronegative ones. Therefore, they are the salts most likely to contain discrete charged ions. However, combinations of the alkaline earth metals (Be, Mg, Ca, Sr, Ba, Ra) and the elements of the sixth column of the periodic table (O, S, Se, Te) also tend to be ionic, and Fig. 2.13 shows that

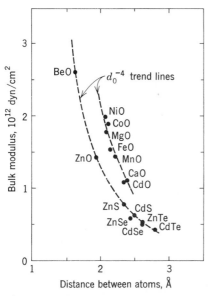

Fig. 2.13 Effect of atomic spacing on the elastic stiffnesses of some oxide crystals.

their bulk moduli correlate with r_0^{-4} trend lines if the structure is held constant. A set of oxides is shown which has the rock salt structure, as well as a set of tetrahedrally bonded crystals, having either the zinc blende or the wurtzite structure.

Rather complex ionic crystals that have single ions or charged atomic groups arranged according to the fluorite structure behave in a similar fashion, as shown in Fig. 2.14.

The cohesion of metal crystals contrasts with salts because the bonding electrons in the simple metals are nearly completely nonlocalized; that is,

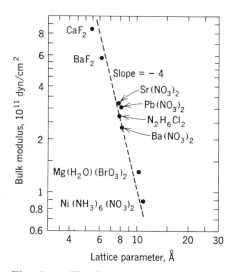

Fig. 2.14 Elastic stiffnesses of several fluorite-type crystals. (*Data from Haussuhl*, 1963.)

they constitute a "gas," and it is the compressibility of this gas that determines the elastic compliance of the metal to a first approximation.

The elastic stiffness of a metal can be estimated from the Sommerfeld free-electron model, but it should be remembered that this oversimplifies the actual problem because it does not take into account the finite sizes of the ion cores of the metal atoms. The equilibrium state of a metal results from a balance between the pressure of the electron gas which wants to expand and thereby reduce its kinetic energy U_k and the electrostatic attraction of the negatively charged gas to the positive ion cores.

The pressure of the gas is given by two-thirds of the kinetic energy per unit volume (Kittel, 1967)

$$P = \frac{2}{3} \frac{U_k}{V} \qquad (2.52)$$

and the kinetic-energy density is related to the product of the electron density N/V and the average kinetic energy per electron which equals three-fifths of the Fermi energy U_f. But the Fermi energy depends on the electron density, so the net result is

$$P \sim \left(\frac{N}{V}\right)^{\frac{5}{3}} \qquad (2.53)$$

which indicates that the pressure decreases with increasing volume for a given number of electrons.

A simplified model for the electrostatic energy lets it be that of a concentrated positive ion immersed at the center of a sphere that has one electronic charge e distributed uniformly through it. The sphere contains an atomic volume and has a radius r_s. The corresponding electrostatic energy is

$$U_e = -\frac{9e^2}{10r_s} \qquad (2.54)$$

This becomes increasingly negative as r_s becomes small, thereby concentrating the distributed electron near the positive core.

A minimum energy exists when $\partial(U_k + U_e)/\partial r_s = 0$, which gives the value of r_s^* when the system is stable:

$$r_s^* = 2.45 \frac{\hbar^2}{e^2 m} = 2.45 r_B = 1.3 \text{ Å}$$

where $r_B =$ Bohr radius. At the equilibrium value of the radius, the bulk modulus is

$$B = -V \frac{\partial P}{\partial V} = \frac{3e^2}{40\pi(r_s^*)^4} \qquad (2.55)$$

which approximately equals the observed values and which indicates that B is proportional to the inverse fourth power of r_s^*. Measured values agree with this as shown in Fig. 2.15.

Complete theories that take actual atomic structures into account have been reviewed by Huntington (1958) and Brooks (1963). Their value to the nonspecialist is limited, however, because they involve numerous assumptions.

The transition metals of the second and third long periods of the periodic table show another correlation that is of considerable interest. In these elements the interatomic spacings are approximately equal, yet their elastic stiffnesses vary by more than a factor of 20. The variations can be correlated if it is assumed that the 5s and 6s electrons have high compliances, as they do according to Eq. (2.55) when their values for r_s^* are inserted in it. They

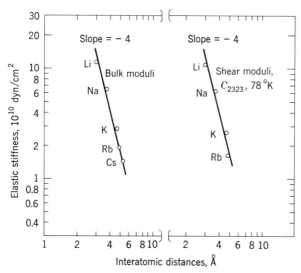

Fig. 2.15 Elastic stiffnesses of alkali metals as they depend on interatomic spacings.

contribute only $\sim$10 percent of the observed stiffness, so most of the stiffness comes from the d-shell electrons, and if the bulk modulus is plotted as a function of the number of d electrons divided by the fourth power of the interatomic spacing, a linear correlation results, as shown in Fig. 2.16. The

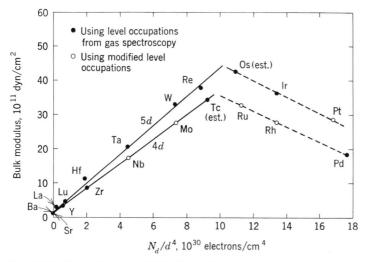

Fig. 2.16 Dependence of the bulk moduli of the second- and third-period transition metals on the number of d electrons per atom divided by the fourth power of the interatomic spacing.

bulk modulus is used here in order to minimize the influence of crystal structure on the correlation.

For the present purposes, covalent crystals can be considered to consist of positive ions with pairs of electrons lying between them and holding them together. This has been demonstrated by experimental x-ray studies (Gotlicher and Wolfel, 1959) and is the basis of a detailed theory (Phillips, 1967). It leads to the immediate conclusion that the bulk modulus should correlate with d^{-4}, as shown in Fig. 2.17 (Keyes, 1962). Changes of volume affect

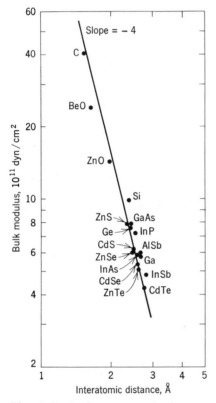

Fig. 2.17 Resistance to elastic compression of covalent tetrahedrally bonded crystals.

only the interatomic distances and not the bond angles, so the bulk modulus is sensitive not to structural patterns but chiefly to spacings. On the other hand, shear changes both bond lengths and angles, with no net volume change. This introduces an additional spacing dependence, so the shear stiffness C_{2323} varies as the fifth inverse power of the spacing, instead of the fourth (Fig. 2.18). This confirms that resistance to bond bending depends sensitively on bond length.

Systematic variations of elastic stiffness can sometimes be used to aid in

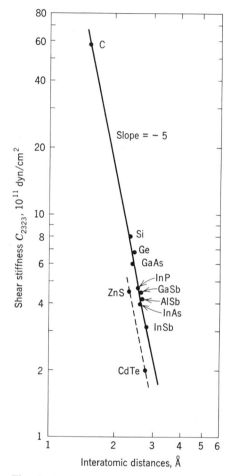

Fig. 2.18 Dependence of shear stiffness on interatomic distances in tetrahedrally bonded crystals.

interpreting bonding mechanisms. This has already been demonstrated by Fig. 2.16, which shows the importance of the d-band electrons in the bonding of the transition metals. Figure 2.19 provides another illustration in the case of the carbides. Because carbide crystals often have the same structural arrangement of their atoms as that of rock salt, some authors have regarded them as ionic crystals. If this were true, the trend line in Fig. 2.19 should have a slope of -4. It clearly does not, and in fact there is little if any dependence of stiffness on atomic spacing, so the bonding in these carbides

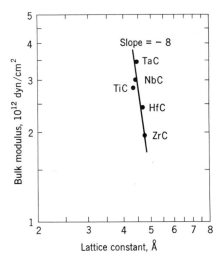

Fig. 2.19 Elastic stiffnesses of some face-centered-cubic carbides as they depend on interatomic spacings.

is akin to that of the transition metals in which the valence factor dominates the size factor.

Finally, molecular crystals that are bound together by dipole-dipole forces (as contrasted with the monopole forces that have been discussed thus far) will be considered briefly. The prototype molecular crystals are the solid rare gases Ne, Ar, Kr, and Xe. Their bulk moduli increase with increasing lattice parameter. This is because the polarizability factor is just as important as the size factor in this case.

According to the molecular theory of cohesion in the rare gases (Kittel, 1967), the bulk modulus is

$$B \simeq \frac{\epsilon}{(r_s^*)^3} \tag{2.56}$$

where ϵ is the pair-potential interaction energy which depends on the atomic polarizability α. Thus a plot of ln (B/α) versus ln r_s^* should yield a line of slope $= -3$, and Fig. 2.20 shows that this is indeed the case. Thus induced

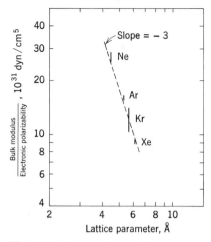

Fig. 2.20 Comparison of theory and experiment for the elastic stiffnesses of the solid rare gases.

dipole interactions can account for the stiffnesses of the rare gas solids (and probably other molecular crystals). Notice that increasing the strengths of permanent or induced dipoles in molecular crystals increases their stiffness. This is done in practice to increase the stiffnesses of polymeric solids.

REFERENCES

Andrade, E. N. da C.: *Proc. Roy. Soc. (London)*, **201**: 439 (1950).

Barlow, W.: *Z. Krist.*, **29**: 433 (1898).

Born, M.: "Atomtheorie des festen Zustandes," B. G. Teubner, Leipzig, 1923.

Brooks, H.: Binding in Metals, *Trans. AIME*, **227**: 546 (1963).

Feynman, R., R. Leighton, and M. Sands: "The Feynman Lectures on Physics," vol. 2, Addison-Wesley Publishing Company, Inc., Reading, Mass., 1964.

Gotlicher, S., and E. Wolfel: *Z. Elektrochem.*, **63**: 891 (1959).

Haussuhl, S.: *Phys. Stat. Solidi.*, **3**: 1072 (1963).

Hooke, R.: "Lectures de Potentia Restitutiva," Published for John Martyn, London (1678).

Huntington, H. B.: The Elastic Constants of Crystals, *Solid State Phys.*, **7** (1958).

Keyes, R. W.: Elastic Properties of Diamond-type Semi-conductors, *J. Appl. Phys.*, **33**: 3371 (1962).

Kittel, C.: "Introduction to Solid State Physics," John Wiley & Sons, Inc., New York, 1967.

Krishnan, K. S., and S. K. Roy: *Proc. Roy. Soc.*, **A210**: 481 (1952).

Mott, N. F., and H. Jones: "The Theory of the Properties of Metals and Alloys," Dover Publications, Inc., New York, 1958.

Nye, J. F.: "Physical Properties of Crystals," Oxford University Press, Fairlawn, N.J., 1957.

Phillips, J. C.: A Posteriori Theory of Covalent Bonding, *Phys. Rev. Letters*, **19**: 415 (1967).

Voigt, W.: *Ann. Phys. Chem. (Wiedemann)*, **31** (1887); **34, 35** (1888); **38** (1889); also, "Lehrbuch der Kristallphysik," B. G. Teubner, Berlin, 1910.

3

CRYSTAL
PLASTICITY

3.1 INTRODUCTION

With its extensive use of the forge, the rolling mill, and machine cutting tools, industrial production depends crucially on the fact that crystals exhibit plasticity. Another area in which crystal plasticity is very important is geology. When crystalline rocks are subjected to combined shear and pressure, they may flow readily, causing them to change their shapes. This leads to the development of rock textures and to movements within the earth such as faulting or, on a large scale, earthquakes. Plasticity is of interest also because it often limits strength. It can do this in a variety of ways; by changing the shape of a structural member, by causing stress concentrations that result in fracture, and by causing structural degradation inside a solid which leads to fracture (fatigue).

Most solids are aggregates of very small crystals which tightly adhere to one another. Deformation of the aggregate requires deformation within the crystals, and therefore can only be understood in terms of crystallographic mechanisms.

The change of shape of a body is not the only consequence of plasticity. For example, localized plastic flow can concentrate the stresses that it relieves at the boundary of the flowed region, and if the concentration is sufficiently severe, fracture will result. Also, internal structural changes accompany the flow process. These changes sometimes have marked effects on the chemical, electrical, and magnetic properties of the materials.

A most remarkable feature of crystal plasticity is the large range of resistance to deformation that crystals possess. If the plastic resistance of a soft crystal such as talc is taken as unity, then the resistance of diamond is about 10^8 times as great at ordinary temperatures. This variation is much greater than the variation of elastic stiffness for these substances, and it is a consequence of the dislocation mechanism of the flow process. Between talc and diamond, metal crystals cover less than half the hardness range, and yet the difference between the plastic resistance of lead and steel crystals is striking. The former can be scratched with the fingernail, while the latter can easily scratch glass.

Plastic resistance depends, first of all, on the strengths of the chemical bonds that exist between adjacent atoms in the structure. Next, it depends on the atomic pattern or crystal structure, because the ease with which dislocations move through a crystal is sensitive to the periodicity of the pattern. In addition, other factors can have a strong influence, an important one being impurities. Crystals that are not pure are usually much more resistant to deformation than those that are pure. In some cases the effects are very large, making impure crystals 10 to 100 times more resistant. This results from disturbance of the periodic crystal structure by the impurities and from local bonds that arise between impurities and their neighbors.

Damaging treatment, such as bombardment with high-energy particles, is another means of disturbing the periodicity of the crystal structure. It also usually raises the plastic resistance of a crystal. Furthermore, the states of aggregation of impurities or of defects induced by radiation damage are important. As isolated point defects they usually have their greatest specific effect, but when aggregated into platelets or particles, they can also be effective.

Hard particles lying on an active glide plane in a crystal act as "keys" that prevent local sliding of the planes, thereby increasing the average plastic resistance. This kind of hardening has had a very important role in the development of technologically useful metallic alloys, beginning with aluminum-copper alloys (duralumin) and now touching almost all structural metals as well as many nonmetallic materials.

The mechanical behavior of some crystals is remarkable because the plastic resistance (flow stress) is almost independent of the rate of deformation (strain rate). This distinguishes such crystals in a very characteristic way from things as high polymers and glasses. When they are subjected to a given stress for a short time, these crystals change their shapes almost as much as if they were stressed for a long time. By contrast, hot glass changes its shape in proportion to the duration of an applied stress and is therefore said to have ideal Newtonian viscosity. Furthermore, the low sensitivity of crystals to strain rate is not always accompanied by low sensitivity to temperature. This is one of the peculiarities of the behavior of dislocations.

Not all crystals are insensitive to strain rate. Some, such as germanium and aluminum oxide, are normally quite brittle and can only be deformed plastically if they are stressed very slowly and uniformly. Their plastic resistance is also quite dependent on temperature.

During plastic flow, the internal structure of a crystal changes, and this causes subsequent flow to become increasingly difficult. This *strain-hardening* phenomenon makes a crystal that has been deformed more difficult to deform a second time. The effect can be so intense that a strong man can easily bend a good crystal of copper in the form of a $\frac{1}{2}$-in.-diameter bar into the shape of a horseshoe once, but he cannot then straighten it out. This is because the plastic resistance of the copper crystal increases at least tenfold during the first bending.

Strain-hardening is most pronounced in crystals that glide simultaneously on two or more intersecting planes, because mutual interference between glide processes on the two planes causes "traffic jams" which strongly inhibit further glide. However, some hardening occurs even for glide on a single set of parallel planes.

Hardening similar to strain-hardening comes about when two or more crystals are connected together to form a polycrystalline aggregate. Straining the aggregate causes each crystal to deform by gliding along its preferred glide planes, but the active planes of the various crystals do not match up, so there is interference between the gliding processes in adjacent crystals and, hence, a large amount of plastic resistance. A process that takes advantage of this phenomenon is the forging of metals. During forging, the large crystals of a cast metal are hammered so as to break them down into small crystallites. This increases the amount of glide interference in the aggregate and makes the metal harder and tougher. The fact that fine-grained rocks tend to be stronger than coarse-grained ones is another phenomenon that is in part caused by glide interference.

For many years no one knew how crystals could be plastic. It seemed that the strong interatomic bonds in them should be elastic up to large strains (say 10 percent), whereupon they should fracture. Then it was realized that the heterogeneous nature of the flow process meant that it took place by the motion of highly localized dislocation lines through a crystal. These dislocations have many fascinating geometric and dynamical properties, which fact has stimulated intensive scientific study of them and has gradually been leading to a molecular theory of the flow process.

3.2 CRYSTALLOGRAPHIC ASPECTS

Two distinctive features of crystal plasticity are (1) that it is *anisotropic* because certain crystallographic planes and directions are favored over others; and (2) that it tends to be highly *heterogeneous* because selected planes (or sets of planes in the case of twinning) become sheared while others remain undeformed. This causes initially smooth external surfaces to become roughened by glide or twin markings (called bands).

These features were first studied and clearly described about a century ago by a mineralogist named Reusch (1867), who applied compressive stresses to rock salt (NaCl) and sylvine (KCl). By means of a polarizing microscope, he observed glide bands within them and on their surfaces. From these he determined the planes {110} and directions ⟨110⟩ along which shear deformation occurred (Fig. 3.1). He also studied twin bands in calcite

Fig. 3.1 Montage of x-ray microscope images of the faces of a salt crystal that has been slightly compressed. The material is LiF, which has the crystal structure of rock salt. The straight, dark lines are glide bands filled with dislocations; the curved lines are subgrain boundaries. The glide planes are {110} and the glide directions ⟨110⟩. Magnification = 20×. (*Courtesy of J. B. Newkirk.*)

(CaCO₃) (Fig. 3.2). Thus the two fundamental modes of plasticity were established, namely,† *translation-gliding* and *twin-gliding*. These names will often be shortened to *glide* and *twinning* in the discussion that follows. Translation-gliding is commonly shortened to "slip" in the English metal-

Fig. 3.2 Twinning in calcite produced by simple compression. The crystal at the right was compressed in a small vise along the diagonal axis that goes from its lower right to its upper left. For comparison, an untwinned calcite crystal is shown at the left.

lurgical literature, but this will not be done here because of the mineralogists' precedence and because it leads to confusion with the "slip-line fields" of the continuum theory of plasticity.

CONSERVATION OF STRUCTURE

Numerous studies of crystal plasticity in mineral crystals followed those of Reusch, and a review of the early work was written by Johnsen (1913). One of the fundamental findings was that crystal structure tends to be preserved during a plastic change of shape. This was not a surprise, since it is consistent with energy minimization (the crystalline form being the minimum energy state for an aggregate of atoms). However, its firm experimental establishment was valuable because the fact that plastic flow does not destroy the crystal structure severely limits the number of ways in which the deformation can occur. Mügge (1898) pointed out that the only two modes of deformation that completely conserve a crystal's structure are translation- and twin-gliding.

† This terminology is due to M. J. Buerger (1930). The German equivalents were *gleitung* and *einfache Schiebung*.

One means for deforming a crystal without destroying its atomic pattern is simply to slide (translate) one part of the crystal over the rest by an amount exactly equal to the size b of the unit pattern, or by some integral multiple thereof (Fig. 3.3). Such translations always restore registry of the

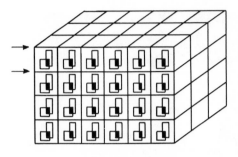

Undeformed crystal

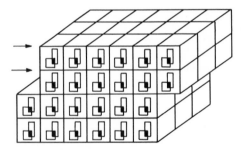

Translation-gliding

Fig. 3.3 Translation-gliding of one unit distance in a crystal. After each unit translation, the crystal structure is restored.

pattern, but out of the large number of geometrically possible translations in a given structure, only those that have a kinetically favorable mechanism will actually take place when a stress is applied. The displacement per active plane is nb, where n is an integer whose value ranges from 1 to more than 10^4 and b is the size of the structural pattern.

The geometric conditions for twin-gliding are somewhat more restrictive than for translation-gliding but still are quite commonly met. Specifically, a plane that initially lacks mirror symmetry, such as ABC in Fig. 3.4, is required. Then twin-gliding can cause the plane to acquire mirror symmetry, as for $A'B'C'$. Since the twin has the same structure as the original matrix (rotated 180° about BC), the structure is thereby conserved. Furthermore, continuity is preserved at the boundary between the two regions. The shear strain is constant in the twinned region, so the displacement of each plane

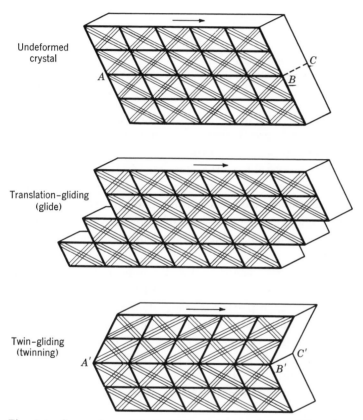

Fig. 3.4   Comparison of twin-gliding and translation-gliding. Both modes of deformation conserve the crystal structure.

increases in proportion to the distance from the reference plane. Notice that the displacement vector in twinning has a definite direction, whereas in translation-gliding it is bidirectional.

In the case of a cleaved calcite rhomb, the external shape becomes twinned when the internal structure does. However, this is not necessarily the case, and Fig. 3.5 illustrates a situation in which the internal structure becomes twinned, but this is not obvious from the external shape change.

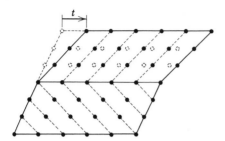

Fig. 3.5   Deformation that generates twin of lattice points without twinned external shape.

Another possibility is for shear to cause a structural transformation rather than a twin, because this can also change the external shape. This is beyond the scope of the present discussion, however.

Only certain planes in a crystal have the symmetry and structure required for twinning. As mentioned above, the symmetry must be such that a simple uniform shear can yield the twinned relationship, and the *structure* must be such as to allow easy rearrangement of the atomic pattern as the shearing process takes place. Since many planes have the required symmetry and yet do not exhibit twinning, it is clear that the actual structure is quite important in determining whether a crystal can twin. Further indication of this is that structures which produce growth twins do not always yield mechanical twins. Prior to twinning, a twin plane has relatively low symmetry, so it is not unexpected that crystals of low overall symmetry tend to twin more profusely than those of higher symmetry. Thus cubic crystals are difficult to twin (e.g., aluminum, copper, silver), and they tend to twin on only one plane (e.g., iron, molybdenum). On the other hand, lower-symmetry metals such as titanium, zirconium, and uranium twin profusely on several different planes.

For both translation- and twin-gliding, the shape change increases the energy of the solid slightly, because new external surfaces are created by the former process and an internal boundary is created by the latter. However, these are usually very small increases compared with the work done during the deformation.

Twin-gliding produces a fixed amount of strain that is determined by the geometric parameters of the structure. Translation-gliding contrasts with this because it can occur repetitively along the same plane to produce very large strains. As an example, Fig. 3.6 shows a zinc crystal that the author stretched 500 percent without destroying its crystal structure and with relatively little internal distortion as determined by x-ray diffraction. Other

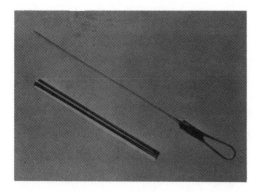

Fig. 3.6 Zinc crystal that was plastically stretched to five times its original length without destroying its crystallinity (only one end is shown). Below the stretched crystal is a piece of the undeformed crystal to indicate the five-fold decrease in thickness that occurred.

crystalline substances, such as aluminum and rock salt, tend to deform simultaneously on intersecting planes. This results in internal distortions and diffuse x-ray diffraction patterns, but the essential crystallinity is nevertheless preserved.

When a crystal deforms by gliding, some atomic planes do not glide at all, some shear one or two unit distances, and some shear hundreds or thousands of unit distances. During twinning, the individual planes shear by a distance that is only a fraction of the structural pattern length, but every successive plane is sheared by this amount. Thus twinning is more homogeneous than gliding.

Both translation- and twin-gliding are geometrically reversible, and in real crystals they are physically reversible under favorable conditions. Geometric reversibility is not necessarily accompanied by thermodynamic reversibility. However, the *paraelasticity* caused by small dislocation motions and the *ferroelasticity* caused by twinning may sometimes be thermodynamically reversible. An example of the latter case is *elastic twinning* in calcite (Klassen-Neklyudova, 1964).

Twin-gliding will not be discussed here in detail. Good reviews are available elsewhere, such as in the reference just above.

GLIDE ELEMENTS

Although many mineral crystals are too brittle at normal temperatures to be deformed in simple compression, they can be deformed without fracture if they are immersed in a fluid such as powdered sulfur that is maintained under pressure during the deformation. In this way, mineralogists (especially Mügge) determined the crystallographic glide elements for a number of crystals (Tertsch, 1949). Later, the work of Ewing and Rosenhain (1899, 1900) and others showed that the plasticity of metal crystals is quite similar to that of mineral ones, and when synthetic metal crystals began to be grown (Andrade, 1914; Carpenter and Elam, 1921), many glide elements for metals were determined. Still later, numerous synthetic nonmetallic crystals became available, so the present list of known glide elements has become quite extensive.

Following early nomenclature, the glide planes and directions for translation-gliding are designated $B(hkl)$ and $b(uvw)$, respectively. For twin-gliding, the planes and directions are $T(hkl)$ and $t(uvw)$. The set of glide elements that operates most easily (requiring the least shear stress) at low ambient temperature is said to be the *primary* set; the next most easily operable is the *secondary* set, etc. The methods used for determining glide elements have been analysis of surface marking, Taylor's analytic method (Taylor and Elam, 1923) for obtaining the strain ellipsoid, and x-ray studies of the lattice rotations that occur during deformation. Table 3.1 lists known glide elements for several structure types. It is not feasible to list all crystals here, but the probability is high that crystals with the same structure will have the same glide elements. An extensive listing of known elements is given by Handin (1966).

A simple rule usually determines the primary glide direction b for a given structure. It is simply the one with the shortest possible translation vector (Fig. 3.7). This is true for nearly all the crystals listed in Table 3.1. Exceptions are compounds with the ordered Cu_3Au structure and elemental mercury.

Of the multitude of planes that contain the glide direction, the one that is most smooth on a molecular scale is usually preferred. This results because geometrically smooth planes are the most widely spaced ones in a structure and, hence, the most weakly bound together. However, this geometrical rule sometimes is not obeyed, and it appears that the selection of the primary glide plane depends on the relative strength of the chemical binding across the plane. Iron crystals (which form one exception to the geometrical rule) glide with almost equal ease on several planes of the $\langle 111 \rangle$ zone. More definitive evidence is provided by the fact that crystals with the same structures (but different chemical constitutions), sometimes have different glide planes (but the same glide directions). Figure 3.8 illustrates this for the rock

TABLE 3.1

CRYSTAL CLASS	STRUCTURE TYPE	SUBSTANCES (examples only; not comprehensive)	OBSERVED TRANSLATION-GLIDE ELEMENTS		CLOSEST-PACKED PLANES	SHORTEST TRANS. DIRECTION
			B	b		
Cubic	Face-centered	Al, Cu, Ag, Au, Pb, v-Iron, Ni, CuAu, α-CuZn AlZn, AlCu	$\{111\}$	$\langle 10\bar{1} \rangle$	$\{111\}$	$\langle 10\bar{1} \rangle$
	Body-centered	α-Iron, W, Mo, Nb, Ta, β-Brass	$\{101\}$ or $\{211\}$	$\langle 111 \rangle$	$\{101\}$	$\langle 111 \rangle$
	Diamond	C, Si, Ge, α-Sn	$\{110\}$	$\langle 1\bar{1}1 \rangle$	$\{110\}$	$\langle 1\bar{1}1 \rangle$
	Rock salt	NaCl plus most alkali halides	$\{110\}$	$\langle 1\bar{1}0 \rangle$	$\{110\}$	$\langle 110 \rangle$
		TiC	$\{111\}$			
		PbTe	$\{001\}$			
	Cesium chloride	MgTl, CsCl, NH$_4$Cl, Br, AuZn, MgTl, TlCl, AuCd	$\{110\}$	$\langle 100 \rangle$	$\{110\}$	$\langle 100 \rangle$
		β-brass	$\{110\}$	$\langle \bar{1}11 \rangle$	$\{110\}$	$\langle 100 \rangle$
	Fluorite	CaF$_2$, BaF$_2$, UO$_2$, ThO$_2$	$\{100\}$	$\langle 110 \rangle$	$\{100\}$	$\langle 110 \rangle$
	Zinc blende	InSb	$\{111\}$	$\langle \bar{1}10 \rangle$	$\{111\}$	$\langle 1\bar{1}0 \rangle$
Tetragonal		β-Sn	$\{110\}\{100\}$	$\langle 001 \rangle$	$\{100\}$	$\langle 001 \rangle$
Hexagonal	Nearly close-packed	Mg, Zn, Cd, Be, Ti	$\{0001\}$	$\langle 11\bar{2}0 \rangle$	$\{0001\}$	$\langle 11\bar{2}0 \rangle$

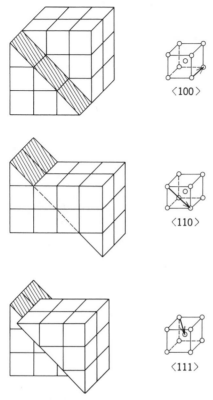

Fig. 3.7 Three possible glide directions in iron. The shortest one ⟨111⟩ is preferred.

salt structure. Both galena (PbS) and rock salt (NaCl) glide in the ⟨110⟩ direction, but whereas galena prefers the close-packed and most widely separated {100} planes, rock salt prefers the more closely spaced {110} planes.

Another feature of translation-gliding is that it is not always bidirectional (Mügge, 1898), that is, the shearing stress that is needed to cause glide in a given direction is not necessarily the same as that needed to cause glide in the opposite direction. The behavior depends on whether or not the glide plane is symmetric about a line that is perpendicular to the glide direction. Asymmetric glide planes have sawtoothed potential-energy structures instead of symmetric hills and valleys. Examples of unidirectional crystals are hydrated barium bromide ($BaBr_2 \cdot 2H_2O$) and potassium chlorate ($KClO_3$). Also, it is expected that a crystal such as aluminum oxide (Al_2O_3) which has

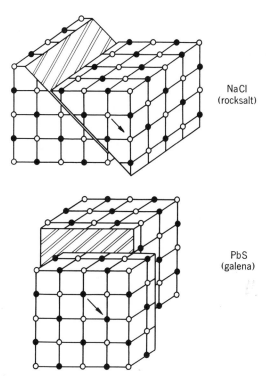

NaCl
(rocksalt)

PbS
(galena)

Fig. 3.8 Translation-gliding in the same direction, but on two different planes, in crystals that have the same structure; rock salt prefers the {110} plane, whereas galena prefers {100}.

a trigonal axis perpendicular to the glide plane (0001) should exhibit microscopic unidirectionality because a translation along any direction of a particular basal plane is not the same as the reverse translation.

A selection criterion for the elements of twin-gliding has been proposed by Jaswon and Dove (1960). Of all possible twinning elements in a structure, they propose that the primary set will be the one that has the least twinning shear associated with it. This rule is true for several monatomic metal crystals, but it has not been tested for a broad variety of structures.

CONSERVATION OF DENSITY

A corollary of the observation that crystal structure does not change during deformation is that the lattice dimensions and density also change very little. Even very severe deformations (especially if the flow is laminar) cause density changes of only several percent.

From the lack of density change it follows that the components of plastic strain are not all independent, because the sum of the principal tensile strains must be zero ($\epsilon_{11} + \epsilon_{22} + \epsilon_{33} = 0 =$ plastic dilation). Therefore, for bidirectional shears there are five independent components out of six, and eight out of nine for unidirectional shears.

HETEROGENEITY OF DEFORMATION

It has already been mentioned that small markings usually appear on one or more surfaces of a crystal when it is deformed (Fig. 3.9). Their orienta-

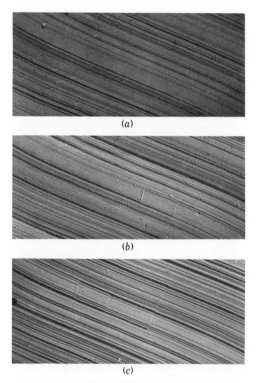

(a)

(b)

(c)

Fig. 3.9 Progressive development of glide lines on the surface of an α-brass crystal: (a) glide shear strain = 0.32; (b) glide shear strain = 0.50; (c) glide shear strain = 0.78. Magnification = 10,000×. (*Courtesy of J. T. Fourie.*)

tions and intensities are related to the crystallographic glide elements. The presence of these sharply defined surface markings (with a distribution of intensities) proves that the flow is heterogeneous. Therefore, the average microscopic strain may be quite different from the average macroscopic strain, so the latter is likely to be a poor measure of the internal state of the strained material. As we shall see later, the instantaneous *strain rate* is a better measure, although the strain is also useful.

The most extreme localization of plastic strain occurs when many units of translation-gliding pass over a single atomically flat plane. The displacement might be as much as 100 atomic spacings, and since the glide-plane spacing is about one atomic unit, the local strain is 100. In contrast, the macroscopic strain that would be produced by such a displacement in a 3-cm-long specimen would be only about 10^{-6}. Thus fluctuations from the average strain can be very large indeed, and this makes it useful to think of the flow process in terms of specific displacements rather than strains. The term *specific displacement* means the displacement per unit length or height.

DEFORMATION MODES THAT DO NOT CONSERVE STRUCTURE

Figure 3.10 shows that a piece of material can acquire a particular macroscopic shape, such as a curved shape, by more than one direction of localized

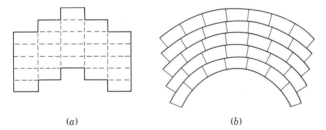

(a)   (b)

Fig. 3.10 Different deformation modes that produce an external curved shape: (a) gliding on vertical planes; (b) gliding on horizontal planes.

shearing. In Fig. 3.10a the shape change is achieved through translation-gliding with complete restoration of the internal structure after the deformation. In Fig. 3.10b the shape change has been produced in a different way, and this has caused internal disregistry of the structure. It should be noted that (a) required that a vertical glide plane be available and that the shapes are different in detail. Oblique glide planes would have produced a mixture of the (a) and (b) types of deformation.

Crystals usually prefer to deform so as to conserve their internal structure, but a general deformation requires six strain components for its description; three tensile and three shear types. As was mentioned previously, however, plastic strain causes no volume change, so the three tensile strains must add up to zero, leaving five independent strains. In an anisotropic crystal (a strongly layered one, for example), five independent glide systems may not be available. If an arbitrary strain is imposed on such a crystal, it has the choice of deforming in a structurally nonconservative way or of fracturing. Whichever of these processes requires the least energy increase will tend to be chosen.

In accordance with the above argument, metal crystals that are imbedded in a polycrystalline aggregate are often found to be bent, twisted, kinked, etc., in addition to being twinned and translated. By using these nonconservative modes, they are able to change their shapes to conform with the deformed crystals that surround them, whereas there might not be enough "pure" modes available for them to deform completely conservatively. The same thing is true of mineral crystals that are constrained by surrounding rocks (Mügge, 1898), and it is common to find bent or twisted mica, calcite, ilmenite, and kyanite crystals embedded in rocks (Fig. 3.11).

Fig. 3.11   A bent beryl crystal embedded in a quartz matrix. (*Courtesy M. L. Kronberg.*)

Essentially the same argument can be made in terms of stress. If the crystal of Fig. 3.10 had only vertical glide planes, it could yield to shear stress applied in the vertical (or horizontal) direction, but not to pure bending moments applied at the ends. On the other hand, the deformation mode in Fig. 3.10*b* allows a crystal to yield to pure bending moments.

Two distinct modes of nonconservative deformation are possible. One consists of bending about an axis that lies in a glide plane (Fig. 3.12) and is called *bend-gliding* after the German *begegleitung*. The other consists of

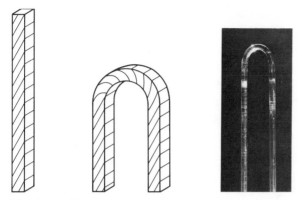

Fig. 3.12 Schematic drawings of bend-gliding plus an example of a sapphire (Al_2O_3) crystal that was bent at a high temperature ($\sim$2000°C).

rotation about the normal to a glide plane (Fig. 3.13) and is called *twist-gliding*.

In both cases, a finite deformation of the crystal (if its dimensions are greater than a few interatomic distances) requires large internal distortions as originally pointed out by Polanyi (1925) for the bending case. For small deformations, the localized distortions may be widely separated by regions

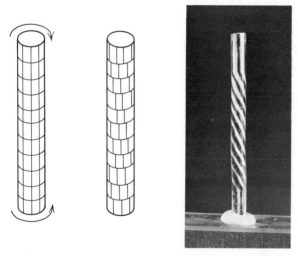

Fig. 3.13 Twist-gliding in a zinc crystal. The crystal was etched with HCl to emphasize the prism planes and then twisted about the hexagonal axis.

of good crystal, but nevertheless they must be present. These atomic mismatches that arise across the active glide planes can be described concisely in terms of dislocations, as will be discussed later. Since the concentration of localized distortions increases with the amount of bending or rotational strain, these nonconservative modes cannot operate indefinitely. Either strain-hardening or fracture will eventually stop them.

Bending and twisting may occur together if a rod-shaped crystal with its glide plane parallel to the rod axis is twisted (Fig. 3.14).

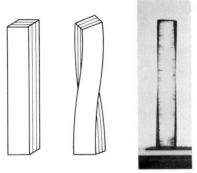

Fig. 3.14 Combined rotation- and bend-gliding in a gypsum crystal $(CaSO_4 \cdot 2H_2O)$ that was twisted about an axis lying in its basal glide plane.

INHOMOGENEOUS DEFORMATION MODES

It has already been pointed out that simple glide and twinning can be inhomogeneously distributed within a piece of deformed crystal. In addition there are certain characteristic types of inhomogeneous deformation that commonly occur. One that leads to interesting shapes and that is quite common is called compressive kinking. It is frequently observed in layered mineral crystals and was first described in detail by Mügge (1898).

An example of a compressive kink for the case of cadmium is shown in Fig. 3.15. Here a crystal with its glide plane inclined slightly with respect to its axis (its trace forms a weak shadow in Fig. 3.15) has been compressed parallel to the axis. This has caused local plastic shearing near the center of the specimen, which has become localized because the glide plane rotated toward a larger angle with respect to the compression axis. This increased the shear stress on the glide plane for a constant compressive load. The condition for this kind of "geometric instability" will be discussed in more detail later.

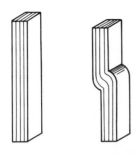

Fig. 3.15 Crystal of cadmium that was kinked by a compression along its axis.

Another inhomogeneous mode that is observed very often on both a macro- and a microscopic scale is that of tensile kink-band formation (Gilman and Read, 1953). Figure 3.16 illustrates this phenomenon which occurs when one part of a crystal is somewhat harder than the remainder. This can be induced by scratches, inclusions, etc.

The sequence of events is that the rate of strain in the hard region lags behind the strain rate in the surrounding material. The planes AA' and BB' bound the hard region. In order for the regions to match in size along the boundaries (that is, to be "compatible"), it is necessary that AA' and BB' bisect the angle between the glide planes inside and outside the central region. This in turn requires that they rotate in opposite directions as the axial stretching proceeds.

Figure 3.16c illustrates the fact that kink bands can exist with a large range of sizes.

In order to emphasize the important restrictions that compatibility requirements place on inhomogeneous plasticity, Fig. 3.17 is presented. If, as in Fig. 3.17b, some material is plastically sheared to a strain $\epsilon_p = \delta/h$, its

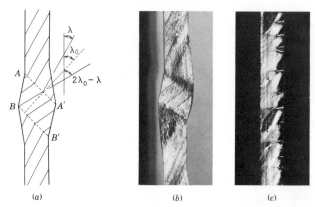

Fig. 3.16 Tensile kink bands in plastically stretched crystals:
(a) schematic drawing showing how central part rotates
clockwise with respect to the axis, while the remainder rotates
counterclockwise; (b) kink band in stretched zinc crystal
(magnification = 4×); (c) series of small kinks plus multi-
tudinous microkinks in a stretched tin crystal.

boundary AB perpendicular to the shear direction becomes lengthened to

$$A'B = h(1 + \epsilon_p^2)^{\frac{1}{2}} \simeq h \left(1 + \frac{\epsilon_p^2}{2}\right)$$

If this is compressed elastically to the length AB, a tensile strain of

$$\epsilon_0 = \frac{A'B}{h} - 1 = \frac{\epsilon_p^2}{2}$$

is introduced. The corresponding stress is $Y\epsilon_p^2/2$ (where Y = Young's
modulus), which becomes very large for small plastic strains. For example,
a plastic strain of 3 percent would produce a stress of about 10^4 psi in steel.
Such large stresses often lead to fracture, as in the case of the "canals" that
open up at twin intersections in calcite because of the strain incompatibility
that occurs (Rose, 1869).

Still another mode of inhomogeneous deformation is called *prismatic
punching* and is illustrated by Fig. 3.18. It commonly occurs in crystals that
have the CsCl structure because the primary glide system is $\{100\}$ $\langle 010 \rangle$, so
the glide directions and planes are mutually orthogonal. This means that
punching can occur when two planes with a common direction become active.

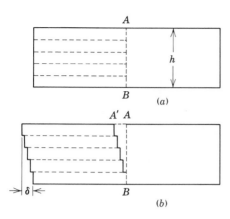

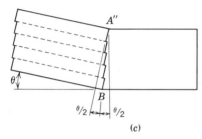

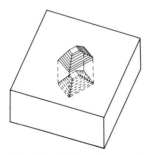

Fig. 3.17 The lamellar nature of glide deformation requires that a shear discontinuity lie in a plane that bisects the misorientation angle between the material on the two sides of the discontinuity AB: (a) undeformed state; (b) material on left strained δ/h (note that $A'B \neq AB$); (c) material on left joined to that on right along matched boundary $A''B$.

Fig. 3.18 Prismatic punching in a thallium bromo-iodide crystal.

3.3 GEOMETRY OF HOMOGENEOUS MACROSCOPIC DEFORMATION

UNIAXIAL EXTENSION AND COMPRESSION VIA TRANSLATION–GLIDING

The fact that crystals flow on preferred planes in special directions causes important changes in the conditions of flow as plastic straining progresses. This was first demonstrated and analyzed in detailed studies of zinc crystals by Mark, Polanyi, and Schmid (1922). The flow conditions change because the active glide plane and direction change their orientation relative to the stress axis during straining. They rotate toward the axis during extension and away from it for compression. The consequences of these rotations are:

1. The external forces do not change orientation as the plastic strain increases, so their resolved components along particular planes and directions do change. Therefore, the shear stress that acts on the initial glide system may decrease while it increases for some other system. It may also increase on a twinning system, thereby causing the deformation mode to change from translation- to twin-gliding.
2. The shear component of an applied axial stress is zero for planes parallel or perpendicular to the stress axis and has its maximum value when the plane is inclined at 45°. If the inclination is initially greater than 45°, plastic tensile strain will cause it to decrease toward 45°, thereby increasing the resolved glide stress without an increase in the applied stress. This leads to negative resistance to plastic flow (or plastic instability) if the rate of strain-hardening is not large enough to overcome the effect. This form of instability is sometimes called *geometric softening*.
3. The amount of local glide strain that is needed to produce a given amount of macroscopic elongation changes with the elongation.

The phenomena listed above will now be described quantitatively. Consider Fig. 3.19, which shows a schematic cylindrical specimen of cross-sectional area A and containing a glide plane defined by its normal $\mathbf{B}$ which makes an angle $90 - \chi$ with respect to the axis $\mathbf{z}$. The initial length along the center line is $\mathbf{L_0}$, and the glide direction is $\mathbf{b}$, which makes an angle λ_0 with respect to the axis. If the specimen is stretched to a length $\mathbf{L}$ the extensional plastic strain is $(L - L_0)/L_0 = \epsilon_p$. The configuration then becomes that shown in Fig. 3.20, where λ has decreased and $90 - \chi$ has increased.

Let $\mathbf{L_0}$ and $\mathbf{L}$ be the initial and final lengths, respectively. $\mathbf{B}$ is a unit vector normal to the glide plane; $\boldsymbol{\delta}$ is the shear displacement vector; $\mathbf{b}$ is a

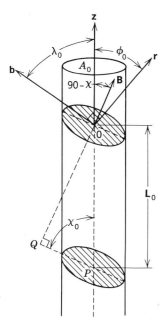

Fig. 3.19 Schematic cylindrical crystal with a single glide system **B**, **b**.

unit vector in the direction of displacement; and ϵ_g is the glide strain.† Then,

$$\mathbf{\delta} = \mathbf{L} - \mathbf{L}_0 = \epsilon_g (\mathbf{L} \cdot \mathbf{B}) \mathbf{b} \tag{3.1}$$

Making the scalar product with **B** on both sides and noting that $\mathbf{b} \cdot \mathbf{B} = 0$ gives

$$\mathbf{B} \cdot \mathbf{L} = \mathbf{B} \cdot \mathbf{L}_0$$

so that

$$\frac{L}{L_0} = \frac{\sin \lambda_0}{\sin \lambda} = \frac{\sin \chi_0}{\sin \chi} \tag{3.2}$$

a well-known result (Schmid and Boas, 1935) that was first used by Mark, Polanyi, and Schmid (1922) and allows the new orientation angles χ, λ to be determined from the extension ratio. Note that as L/L_0 increases, χ and λ decrease.

To find the amount of displacement (3.1) is squared:

$$\mathbf{L} \cdot \mathbf{L} = L^2 = L_0^2 (1 + 2\epsilon_g \sin \chi_0 \cos \lambda_0 + \epsilon_g^2 \sin^2 \chi_0) \tag{3.3}$$

† Note that the tensor strain component would be $\epsilon_g = \frac{1}{2}\epsilon_g$ where ϵ_g is the "engineering glide strain."

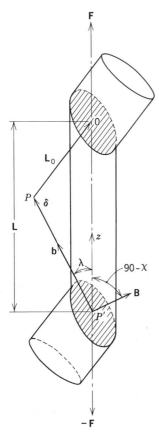

Fig. 3.20 Same as Fig. 3.19
after some plastic deformation
has occurred.

Let $e = L/L_0$ be the extension ratio, and solve for ϵ_g:

$$\delta = \epsilon_g = \frac{1}{\sin \chi_0} [(e^2 - \sin^2 \lambda_0)^{\frac{1}{2}} - \cos \lambda_0] \tag{3.4}$$

which can be rewritten by making use of (3.2):

$$\epsilon_g = \frac{\sin \lambda_0}{\sin \chi_0} (\cos \lambda - \cos \lambda_0) \tag{3.5}$$

Equations (3.3) and (3.4) mean that the amount of glide strain caused by a given amount of extension depends on the initial orientation angles χ_0, λ_0.

Differentiation of (3.3), yields the variation of the extensional strain with the glide strain (since $\Delta\epsilon = \Delta e$):

$$\frac{\Delta\epsilon}{\Delta\epsilon_g} = \sin \chi_0 \cos \lambda \tag{3.6}$$

Therefore, as the elongation increases so λ decreases, the axial strain produced by a given glide strain increases.

It may be seen in Fig. 3.20 that the shape of the cross section changes during extension. This may also be calculated as shown by Von Goeler and Sachs (1927).

In compression the situation simply reverses in principle, but in practice the specimens used for compression are squat and are loaded between rigid plates which restrain rotations at the ends. This changes the geometrical relations (Schmid and Boas, 1935).

For some crystals, especially cubic ones, the structure contains two or more equivalent glide systems (plane plus direction). Then, if two or more of these systems are equally stressed, they will be simultaneously active. The case of two systems has been analyzed by Von Goeler and Sachs (1927); see also, Ewald, Poeschl, and Prandtl (1936). Let the two systems be identified by the unit-vector pairs $(\mathbf{B},\mathbf{b})$ and $(\mathbf{B}',\mathbf{b}')$. Then there are three possibilities:

1. $\mathbf{b} \neq \mathbf{b}'$ $\mathbf{B} = \mathbf{B}$ one plane—two directions
2. $\mathbf{b} = \mathbf{b}'$ $\mathbf{B} \neq \mathbf{B}'$ two planes—one direction
3. $\mathbf{b} \neq \mathbf{b}'$ $\mathbf{B} \neq \mathbf{B}'$ two planes—two directions

The first two cases are relatively simple. For the first one, a resultant unit vector $\mathbf{b}_r$ can be defined:

$$\mathbf{b}_r = \frac{\mathbf{b} + \mathbf{b}'}{|\mathbf{b} + \mathbf{b}'|} \tag{3.7}$$

Then, if λ is the angle between $\mathbf{b}_r$ and the z axis, Eq. (3.2) holds. In the second case, an "effective glide plane" can be chosen such that it is symmetric with respect to the two glide planes. Then the behavior of the effective system obeys the relations given above.

In the third case, by analogy with Eq. (3.1),

$$d\mathbf{L} = d\epsilon_g(\mathbf{L} \cdot \mathbf{B})\mathbf{b} + d\epsilon_g'(\mathbf{L} \cdot \mathbf{B}')\mathbf{b}' \tag{3.8}$$

but $\mathbf{L}$ is symmetric with respect to both $\mathbf{B}$ and $\mathbf{B}'$ and $\mathbf{b}$ and $\mathbf{b}'$, so that

$$\mathbf{L} \cdot \mathbf{b} = \mathbf{L} \cdot \mathbf{b}'$$

and

$$\mathbf{L} \cdot \mathbf{B} = \mathbf{L} \cdot \mathbf{B}'$$

with

$$d\epsilon_g = d\epsilon_g'$$

so (3.8) becomes

$$d\mathbf{L} = d\epsilon_g |\mathbf{b} + \mathbf{b}'|\mathbf{b}_r(\mathbf{L} \cdot \mathbf{B}) \tag{3.9}$$

but this does not correspond to simple glide because $\mathbf{b}_r \cdot \mathbf{B} \neq 0$. Therefore, a resultant glide plane is defined by $\mathbf{B}_r = $ unit-vector normal

$$\mathbf{B}_r = \frac{\mathbf{B} + \mathbf{B}'}{|\mathbf{B} + \mathbf{B}'|} \tag{3.10}$$

and a mean glide strain $\bar{\epsilon}_g$ is defined. Then

$$\bar{\epsilon}_g = \frac{\mathbf{L} - \mathbf{L}_0}{\mathbf{b}_r(\mathbf{L} \cdot \mathbf{B}_r)}$$

so Eq. (8.9) becomes

$$d\mathbf{L} = \tfrac{1}{2}d\bar{\epsilon}_g\, \mathbf{b}_r(\mathbf{L} \cdot \mathbf{B}_r) \tag{3.11}$$

where $\mathbf{b}_r \cdot \mathbf{B}_r \neq 0$.

Multiplying (3.11) by $\mathbf{b}_r$ gives

$$d\mathbf{L} \times \mathbf{b}_r = 0$$

or

$$\mathbf{L} \times \mathbf{b}_r = \text{const} = \mathbf{L}_0 \times \mathbf{b}_r$$

Then, if θ is the angle between the rod axis and $\mathbf{b}_r$, the extension ratio becomes

$$e = \frac{L}{L_0} = \frac{\sin \theta_0}{\sin \theta} \tag{3.12}$$

These various relations have been extensively studied and verified in a variety of crystals. Summary descriptions of the experimental work may be found in the books by Elam (1935) and Schmid and Boas (1935).

BEND–GLIDING

After plastic bending, a crystal contains bent lamellae, so the internal structure as well as the external shape is changed; that is, the structure is not conserved (Mügge, 1898). The geometry was first analyzed by C. D. West (1955), and the analysis was developed by Nye (1953). It has been confirmed by West for bent sapphire crystals, for zinc by Gilman (1955), and most carefully by Bilby and Smith (1956). The simplest case is that of a crystal with a single active glide system being bent about an axis that lies in the glide plane perpendicular to the glide direction. Thus the bending axis would be the vector **r** in Fig. 3.19. Because all the deformation occurs on parallel planes, the lamellae between the glide planes become curved sheets. If the bending is done about an axis that is not perpendicular to a glide direction, the glide lamellae become doubly curved.

Zinc crystals are particularly suited to studies of glide-plane curvature because short glide lines appear on the external surfaces of previously polished specimens and because bent crystals can be cleaved at low temperatures along the glide planes to reveal their curved shapes in toto. An example of this is shown in Fig. 3.21, along with a schematic drawing of the glide-plane shapes. Furthermore, the plastic flow stress is so small compared with the elastic modulus of zinc that very little change of the shape occurs after the load is removed.

A detailed description of the geometric situation is shown in Fig. 3.22. Here, a crystal of thickness $2h_0$ has been bent to a curvature radius R. Since the flow is lamellar, the vertical distance between equivalent points on adjacent glide planes must remain constant. Also, since the residual elastic stresses are negligible, there must be no change in volume, and finally, the glide-plane shape must be consistent with the external shape change. These conditions are satisfied if the glide planes are involutes of a circular evolute. The formal theory which covers this and many other cases has been developed and reviewed by Bilby (1960), but study of Fig. 3.22 should convince the reader that the conditions are satisfied.

Since they are involutes, the glide-plane traces are generated by "a pencil tied to the end of a string that is unwrapped from the evolute." The length of the "string" is always equal to the arc $a\psi$.

If a is the radius of the evolute and a point on it lies at (a,ψ), then the (x,y) coordinates of a point P on the involute are

$$x = a(\cos\psi + \psi\sin\psi)$$
$$y = a(\sin\psi + \psi\cos\psi)$$

$$(3.13)$$

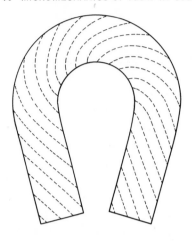

Fig. 3.21   Zinc crystal bent about an
axis lying parallel to its glide planes:
(a) schematic drawing; (b) actual bent
and cleaved crystal.

which are the parametric equations of the involute. When the point P lies
on the neutral line where no strain has occurred, the orientation angle of
the glide plane has its initial value χ_0. The radial vector to P which is designated r becomes equal to R and is perpendicular to the neutral line. The
length of the "string" which generates the involute equals $a\psi$ and is always
perpendicular to a radius of the evolute. Therefore, $\chi_0 = 90 - \psi$, and

$$a = R \sin \chi_0 \tag{3.14}$$

so that a can be found from the initial orientation and the amount of bending. Similarly, the curvature radius of the glide planes is

$$c = a\psi \tag{3.15}$$

and ψ is obtained in terms of the given quantities by setting

$$r^2 = x^2 + y^2$$

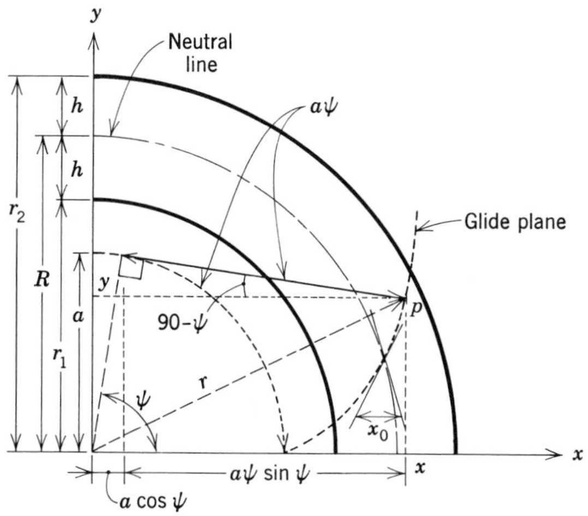

Fig. 3.22   Detailed geometry of a plastically bent crystal.

so, upon substituting (3.13),

$$\psi^2 = \frac{r^2 - a^2}{a^2}$$

From (3.15),

$$c^2 = r^2 - a^2 = r^2 - R^2 \sin^2 \chi_0 \tag{3.16}$$

and the glide-plane curvature is determined for any radial position.

When $r = a$, the curvature $K = 1/c$ becomes infinite, and the glide planes at $r = a$ become perpendicular to the free surface. This last may be seen by inspection and also follows from the extension rule [Eq. (3.2)]:

$$\sin \chi = \frac{L_0 \sin \chi_0}{L} = \frac{R \sin \chi_0}{r} = \frac{a}{r}$$

so when $r = a$, $\chi = 90°$.

The critical value of $R = R^*$ that causes c to become zero may be found by applying the rule that plastic flow does not change the volume. The volume of a fiber element before deformation is $2\pi hR$, and after the deformation which makes $a = r_1$, the fiber's volume becomes $\pi(R^2 - a^2)$, assuming that the neutral plane is fixed. Then

$$a^2 = R(R - 2h)$$

but $a = R \sin \chi_0$, so

$$R_{\min} = \frac{2h}{\cos^2 \chi_0} \tag{3.17}$$

and, for example, if $\chi_0 = 45°$, we see that $R_{\min} = 4r_0$. Thus h_0 should be as small as possible if sharply bent glide planes are to be produced.

A crystal can only be bent further than the limit given by Eq. (3.17) by a combination of extension and bending accompanied by inward movement of the neutral plane. Nye (1953) has calculated the resulting shape changes.

The results of Bilby's and Smith's study of experimentally bent zinc crystals are summarized in Fig. 3.23 and act as an excellent confirmation of the theory.

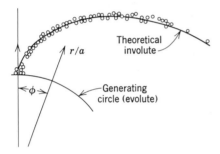

Fig. 3.23 Comparison of points lying along glide planes of experimentally bent zinc crystals with theory. (*After Bilby and Smith.*)

TWIST–GLIDING

It was shown in Fig. 3.13 that crystals may be twisted about the normals to their glide planes. Such deformation can be described quite simply in terms of the rotation angle of one end of the specimen with respect to the other. However, it is clear that the internal structure will be changed unless the rotation angle corresponds to a symmetry element and is localized to a single plane. Otherwise the lattice points must be disarrayed. Thus it is expected that this mode will be favored by high symmetry about the glide-plane normal, because this will minimize the disarrangement.

MIXED MODES

When an arbitrary but uniform set of tractions (tensions, compressions, bending moment, or torques) is applied to a crystal, it will deform by means of some combination of the above three modes. It is especially difficult, of course, for an anisotropic crystal to respond to applied forces using only a single deformation mode. For example, the response of a gypsum crystal that was subjected to a torque is shown in Fig. 3.14 to be a simultaneous combination of bend- and rotation-gliding. This combination of modes has been studied in zinc in some detail by Whapham and Wilman (1956) and analyzed by Bilby and Gardner (1958). Also, a careful study of calcite has recently been completed by Borg and Handin (1967).

TWIN-GLIDING

The shape changes that are caused by twinning were first described in terms of the strain ellipsoid by Liebisch (1887). From Fig. 3.4 it may be seen that, from the viewpoint of the shape change that it causes, twinning consists of a simple shear deformation. Therefore, one axis (perpendicular to the twinning direction) remains unchanged by the deformation.

The fact that the external shape change is a simple shear does not mean necessarily that the internal atomic movements consist only of shear displacements. Even in the case of calcite, the centers of the carbonate ions undergo shear displacements, but in addition the ions must rotate 60° about the normal to the plane of the oxygen atoms. In other structures, more complex "shuffles" may be required in addition to shear displacements.

If the strain-state of the undeformed material is represented by a sphere of unit radius, then a general deformation converts this into an ellipsoid with the directions of the principle strains parallel to the ellipsoidal axes. In the special case of simple shear, the axis of the sphere that lies perpendicular to the twinning direction is one axis of the ellipsoid. Therefore, the geometry may be projected into two dimensions on the plane that is normal to this axis, as shown in Fig. 3.24.

One feature of the deformation is that there are two planes that have the same dimensions after the deformation as before it. The first is the twinning plane defined by the normal vector $\mathbf{T}$. It contains the twinning direction $\mathbf{t}$, so the displacements that are defined by the shear δ occur parallel to it, leaving its position and shape undisturbed. It is conventionally designated K_1. The deformation ellipse intersects the undeformed circle at point C, so this defines another undeformed plane (whose trace is OC). Its initial position is OA, so in this case the position does change during the deformation. The second undeformed plane is conventionally designated K_2.

The angle between K_1 and K_2 is 2ϕ, so its tangent is given by

$$\tan 2\phi = \frac{2}{\delta} \tag{3.18}$$

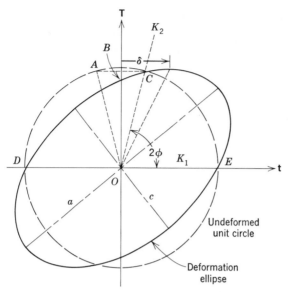

Fig. 3.24   Diagram of the twinning strain ellipse and its
relation to the crystallographic twinning elements.

and the relation between K_1 and K_2 is determined by the magnitude of the
twinning strain $\epsilon_t = \delta$.

Equation (3.3) may be used to determine the positions of the directions
that do not change in length during the deformation. For these directions
the elongation is unity, so Eq. (3.3) becomes

$$1 = 1 + 2\epsilon_t \cos \lambda' \sin \chi' + \epsilon_t^2 \sin^2 \chi' \tag{3.19}$$

where λ' is the angle between the given direction and $\mathbf{t}$ and $(90 - \chi')$ is the
angle between the direction and $\mathbf{T}$.

The solutions of Eq. (3.19) are

$$\sin \chi' = 0$$
$$\frac{\sin \chi'}{\cos \lambda'} = \frac{-2}{\epsilon_t} = - \tan 2\phi \tag{3.20}$$

Here, the second equation defines the second undistorted plane in its initial position. The first equation defines the twinning plane. Together they
divide orientations that contract as a result of twinning from those that
extend. For the special case of $\chi' = \lambda'$, this may be seen quite readily in
Fig. 3.24.

Suppose that a rod-shaped specimen is oriented such that its axis lies somewhere within the segment AOD initially. Then, after twinning has taken place, it will lie somewhere along the segment COD of the deformation ellipse and hence will have been shortened. On the other hand, if it lies within AOE of the circle initially, it will afterward lie within COE of the ellipse and will have been lengthened.

The above statements apply to completely twinned rods. However, as Frank and Thompson (1955) have pointed out, twinning typically begins in a stressed rod as a thin lamella which then thickens. Therefore, rod axes that lie initially in the segment AOB of the circle of Fig. 3.24 do not lengthen when a thin lamella forms; instead, they shorten (but eventually lengthen as the lamella thickens).

One consequence of the results just above is that rods that can shorten during twinning will respond to an applied compression but will not twin if tension is applied along their axes. Rods that are oriented so that twinning will lengthen them will respond to applied tension but not to compression.

A factor that limits the usefulness of twinning as a deformation mode is that the twinning strain is a fixed quantity and is typically small. Therefore, the amount of longitudinal strain that can occur through twinning is small. Its magnitude can be calculated as follows:

The extremal points on the strain ellipsoid are at the ends of its major and minor axes. These are labeled a and c in Fig. 3.24. To find their lengths in terms of the glide strain, the variation of the elongation with orientation angle $\partial e/\partial \chi$ is determined and set equal to zero. The extremal points must lie in the shear plane so $\chi' = \lambda'$. Differentiation of Eq. (3.3) yields

$$\frac{\partial e}{\partial \chi} = \frac{\epsilon_t(\cos^2 \chi - \sin^2 \chi) + \epsilon_t^2 \sin \chi \cos \chi}{[1 + 2\epsilon_t \cos \chi \sin \chi + \epsilon_t^2 \sin^2 \chi]^{\frac{1}{2}}} \tag{3.21}$$

and the numerator must be zero at the extremal positions. Hence, if χ^* signifies the orientation angle for the extremals,

$$(\cos^2 \chi^* - \sin^2 \chi^*) + \epsilon_t(\sin \chi^* \cos \chi^*) = 0 \tag{3.22}$$

and after some trigonometric substitutions,

$$\tan^2 \chi^* - \epsilon_t \tan \chi^* - 1 = 0 \tag{3.23}$$

which has the roots

$$\tan \chi^* = \frac{\epsilon_t \pm \sqrt{\epsilon_t^2 + 4}}{2} \tag{3.24}$$

Equation (3.3) may be written in the form

$$e_{\max} = \tan \chi^* \tag{3.25}$$

or

$$e_{\max} = \pm \frac{\epsilon_t}{2} + \sqrt{\frac{\epsilon_t^2}{4} + 1} \tag{3.26}$$

Since ϵ_t is always less than one for twinning, the maximum obtainable elongation is always small. In some cases this can be enhanced by multiple twinning, but it can never be large. As a result, the importance of twinning as a deformation mode is only modest, and most of the subsequent discussion will be concerned with translation-gliding. For a thorough discussion of twinning, the book edited by Klassen-Neklyudova (1964) is recommended.

3.4 RESPONSE TO STRESS

SCHMID'S LAW (CRITICAL STRESS FOR TRANSLATION–GLIDING)

Experiments have shown that only very small changes of volume accompany large changes of shape during the deformation of metal crystals. Therefore, dilatational or tensile stresses cannot do much plastic work, and they are not expected to have large direct effects on plasticity. Shear stresses may be expected to be of primary importance.

Two kinds of testing have shown that crystals obey quite accurately a critical shear-stress criterion for plastic flow. One kind consists of measuring the yield stresses of identical crystals with and without superposed hydrostatic pressures (Polanyi and Schmid, 1923). These have most recently been performed by Haasen and Lawson (1958), who found that pressures up to about 5,000 atm have very little effect on the yield stresses of Al, Cu, and Ni crystals, although such pressures do affect strain-hardening.

Nonmetallic crystals are more sensitive to pressure than metallic ones, but pressure rarely if ever has a first-order effect (Hanafee and Radcliffe, 1967). Even crystals embedded in rocks and subjected to large pressures deep inside the earth are apparently able to flow readily. It is this insensitivity to pressure that allows some solids such as BN, MoS_2, and graphite to make good lubricants when bearing loads are high.

The idea that shear stress is the dominant driving force for plastic flow has also been tested by performing tensile tests on crystals that had various orientations with respect to the tension axis. The stress component that does the most work during plastic flow and therefore might be expected to be most important is the shear stress that acts on the preferred crystallographic plane, and in the preferred direction. This is the analog of the Tresca or maximum shear-stress criterion for plastic flow in polycrystalline bodies.

To interpret a series of tensile tests, a relation between the applied tension (or compression) and the shear-stress component described above is needed. Referring to Fig. 3.19, if A_0 is the cross-sectional area of the specimen and P is the applied load, then P/A_0 is the axial stress σ_a and the component of the load that acts in the glide direction is $P \cos \lambda$. Since the glide-plane area is $A_0/\cos \phi$, the resolved shear stress is

$$\sigma_s = \sigma_a \cos \phi \cos \lambda \tag{3.27}$$

From this, if rapid plastic flow begins when σ_s reaches a critical magnitude, the tensile stress σ_a^*, at which rapid flow should begin, is

$$\sigma_a^* = \frac{\sigma_s^*}{\cos \phi \cos \lambda} \tag{3.28}$$

Measurements are compared with this relation in Fig. 3.25 for the cases of zinc and cadmium, and it may be seen that the data follow the relation very well. This means that the stress component acting normal to the glide planes has little or no effect on the rate of plastic flow for a large range of values as indicated by the following paragraph.

The normal stress σ_n that acts in Fig. 3.19 is given by

$$\sigma_n = \sigma_a \cos^2 \phi \tag{3.29}$$

so the ratio of normal to shear stress is

$$\frac{\sigma_n}{\sigma_s} = \frac{\cos \phi}{\cos \lambda} \tag{3.30}$$

and for the most symmetric case when $\pi/2 - \phi = \lambda$, so that $\cos \phi = \sin \lambda$; and for a range of λ from 5 to 85°, this ratio varies from 0.087 to 11.4, or a factor of about 130. In the experiments of Andrade and Roscoe (1937), it was found that σ_s^* was within 2 percent of 55.5 g/mm² for σ_n lying between 10 and 215 g/mm².

This behavior is not restricted to metal crystals, but is quite general, as indicated in Fig. 3.26 by some recent data for the molecular crystal anthracene (Robinson and Scott, 1967).

The results just described are summarized in "Schmid's Law of the Critical Resolved Shear Stress" (Schmid and Boas, 1935). This law states that a glide system begins to operate when the shear stress on it reaches a critical

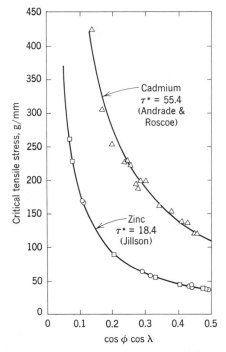

Fig. 3.25 Tensile stresses required to cause plastic flow in Zn and Cd crystals as a function of their orientation angles. (*Data from Andrade and Roscoe, 1937, and from Jillson, 1950.*)

value that is independent of the normal stress. Its corollary is that a crystal which contains crystallographically equivalent glide systems will begin to flow on the system which has the maximum resolved shear stress acting on it.

There have been many interpretations of the Schmid law in the past. It is now understood to be associated with the stress that is needed to overcome the viscous drag that resists the motions of dislocation lines in crystals. This drag depends sensitively on stress for the case of metals, which accounts for the apparently sudden onset of fast flow when a critical stress is reached. The process is more gradual in other crystals, however, with germanium and silicon being good examples. Therefore, the rate of flow can have an important modifying effect on the stress needed for fast flow, and it seems likely that normal stress will also modify the behavior significantly in some cases. Thus, the Schmid law should be viewed as a consequence of more general laws, and it should be expected to be valid in some cases, but not all.

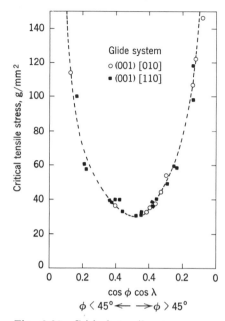

Fig. 3.26 Critical tensile stresses required for the onset of rapid plastic deformation at 298°K in polished anthracene crystals. The dashed curve is calculated from Schmid's law, using a critical shear stress of 14 g/mm². (*Data from Robinson and Scott.*)

CRITICAL STRESS FOR TWIN–GLIDING

Numerous attempts have been made to define a stress criterion for twinning, but no simple criterion has been found (e.g., see review by Cahn, 1954). Unlike translation-gliding, there appears to be little viscous resistance to the propagation of a twin, so no "internal force" controls the rate of propagation. Thus, a characteristic stress for propagation does not appear to exist. The initial nucleation of a twin would depend critically on the applied stress if the process were carried out under homogeneous conditions. In practice, however, various stress concentrators are usually present, making the nucleation process heterogeneous. An exception is the elastic twinning process that was previously mentioned.

RESOLVED SHEAR STRESS CAUSED BY AN ARBITRARY STRESS FIELD

If a set of applied tractions creates a tensor stress field σ_{ij} at a point x_i in a cartesian coordinate system (i,j = 1, 2, or 3), the problem is to specify the stress component acting on an arbitrary glide (or twinning) plane defined by the unit normal vector $\hat{\mathbf{B}}_i$, and in a shear direction that is defined by the unit vector $\hat{\mathbf{b}}_i$ and lies in the plane so that $\hat{\mathbf{b}} \cdot \hat{\mathbf{B}} = 0$.

The stress vector acting on a unit element of the plane is (Feynman, Leighton, and Sands, 1964)

$$\sigma_{in} = \sigma_{ij}\hat{\mathbf{B}}_j$$

in the repeated suffix notation,† so the component of the stress that acts normal to the plane is

$$\sigma_n = \sigma_{ij}\hat{\mathbf{B}}_i\hat{\mathbf{b}}_j \tag{3.31}$$

and the shear-stress component that acts parallel to the plane in the direction $\hat{\mathbf{b}}_i$ is

$$\sigma_s = \sigma_{ij}\hat{\mathbf{B}}_i\hat{\mathbf{B}}_j \tag{3.32}$$

which is consistent with (3.27) above. Also, it may be seen that if σ_{ij} is hydrostatic pressure, so that

$$\sigma_{ij} = \tfrac{1}{3}\sigma_{kk}\delta_{ij} \qquad \delta_{ij} = \text{Kronecker delta} \tag{3.33}$$

and $\sigma_{11} = \sigma_{22} = \sigma_{33}$, then the shear stress becomes

$$\sigma_s = \frac{\delta_{kk}}{3}(\hat{\mathbf{b}} \cdot \hat{\mathbf{B}}) = 0 \tag{3.34}$$

so pressure will not cause flow.

GEOMETRIC PLASTIC INSTABILITY OF CRYSTALS

Because of glide-plane rotation during uniaxial plastic strain for certain initial orientations, the resolved shear stress will increase with strain, although the applied tension or compression remains constant. This can lead to sharp plastic yielding in tension as described by Schmid and Boas (1950) for magnesium and to plastic buckling in the compression of zinc crystals as described by Gilman (1954).

Consider a crystal that requires a certain shear stress σ_s acting on its primary glide element to make it flow. This shear-stress component is created by an applied tension and is in equilibrium with it. The material is hardened

---

† When a suffix appears twice in a given term, summation is to be carried out over that suffix. For example, $X_{ji}Y_i = X_{j1}Y_1 + X_{j2}Y_2 + X_{j3}Y_3$.

by plastic strain, so the flow stress changes with strain ϵ; that is,

$$\frac{\partial \sigma_s}{\partial \epsilon} > 0 \tag{3.35}$$

which is necessary for the flow to be microscopically stable. Let the extension ratio be $e = L/L_0$ and find the variation of the tensile stress with extension, which must be positive ($\partial \sigma / \partial \epsilon > 0$) for macroscopic stability in tension, or negative in compression.

We begin by replacing $\cos \phi$ by $\sin \chi$ (where $\chi = 90 = \cos \phi$) in Eq. (3.27) and differentiating:

$$\frac{d\sigma_s}{de} = \frac{d\sigma}{de} \sin \chi \cos \lambda + \sigma_a \cos \lambda \cos \chi \frac{d\chi}{de} - \sigma \sin \lambda \sin \chi \frac{d\lambda}{de} \tag{3.36}$$

From Eq. (3.2) of the previous section we have

$$\frac{d\chi}{de} = \frac{- \sin \chi}{e^2 \cos \chi} \quad \text{and} \quad \frac{d\lambda}{de} = \frac{- \sin \lambda}{e^2 \cos \lambda} \tag{3.37}$$

and we know that

$$\frac{d\sigma_s}{de} = \frac{\partial \sigma_s}{\partial \epsilon} \frac{d\epsilon}{de} = \frac{\partial \sigma_s}{\partial \epsilon} \frac{1}{\sin \lambda \cos \chi} \tag{3.38}$$

so, after substituting into (3.36) and rearranging,

$$\frac{d\sigma}{de} = \frac{\partial \sigma_s}{\partial \epsilon} \frac{1}{\sin^2 \chi \cos^2 \lambda} + \frac{\sigma_s}{e^2} \frac{\cos^2 \lambda - \sin^2 \lambda}{\cos^2 \lambda} \tag{3.39}$$

and if $\partial \sigma_s / \partial \epsilon$ is positive and large enough, $d\sigma/de$ will always be positive. However, if the rate of strain-hardening is small (say zero), the condition for $d\sigma/de$ to be positive is that

$$\cos^2 \lambda > \sin^2 \lambda \tag{3.40}$$

or, for stable flow,

$$\lambda < \frac{\pi}{4} \quad \text{tension}$$

$$\lambda > \frac{\pi}{4} \quad \text{compression}$$

CONSEQUENCES OF INHOMOGENEOUS PLASTIC STRAINS

It is difficult to accurately characterize the stresses in a crystal that is flowing plastically, because the deformation is so inhomogeneous. Figure 3.27 illustrates this by showing the distribution of plastic strain in a bent crystal of

Fig. 3.27 Lithium fluoride crystal that was slightly bent plastically about the normal to the plane of the photograph and then etched to reveal dislocation lines. The individual etch pits cannot be seen here, but bands of them appear white. Magnification $\sim 10\times$.

lithium fluoride. It may be seen that the boundaries of the plastic zone are made ragged by spike-shaped plastic bands that cross them. After the bands have crossed the neutral plane, it also becomes ill-defined.

Similar situations develop in plastic torsion, indentation, and other inhomogeneous geometries. The stresses are nonuniform both on a macroscale as one passes from the plastic to the elastic zone and on a microscale, where they tend to be concentrated at the tips of plastic bands. Two types of shear concentration are possible: one in which the shear displacement is perpendicular to the line that defines the tip of the band, and the other in which the displacement is parallel to the band tip. In a fully plastic material the maximum shear stress at the tip of such a band cannot exceed the flow stress (approximately). However, under dynamic conditions, or where the material outside the band behaves elastically, large stresses can develop as suggested by Taylor (1928) and analyzed by Starr (1928). The case of shear parallel to the band tip is the more easily analyzed, and this has been done by McClintock (1958) and by Bilby, Cottrell, and Swinden (1963).

Geometric offsets are produced when glide bands intersect free surfaces and they are a source of stress concentration. Their stress-concentration factors have been studied by Marsh (1963), and some of his results are given in Fig. 3.28, where the stress-concentration factor is shown as a function of $(h/r)^{\frac{1}{2}}$, h being the offset height, and r the radius at the step root. Data for steps with flank angles of 90 and 45° are given, the latter being the best model of a glide offset. For comparison, the stress-concentration factors for cracks are given (here h = half crack length, and r = root radius). It is clear that the stresses near the surface of a plastically deformed crystal or near a grain boundary will fluctuate substantially and change during the flow process. Therefore, the internal structures of crystals change during deformation, as well as their stress environments and the microtopographies of their surfaces.

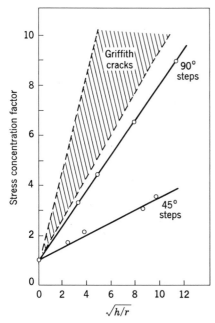

Fig. 3.28 Stress-concentration factors for surface steps as they depend on the ratio of the step height h to the root radius r. Note that steps with 90° flank angles are nearly as severe stress concentrators as Griffith cracks. (*After Marsh.*)

REFERENCES

GENERAL

"Die Festigkeitserscheinungen der Kristalle," H. Tertsch, Springer-Verlag OHG, Vienna, 1949.

"Distortion of Metal Crystals," C. F. Elam, Clarendon Press, Oxford, 1935.

"Handbuch der Physik," "Crystal Physics II," A. Seeger in S. Flugge (ed.), vol. VII 2, Springer-Verlag OHG, Berlin, 1958.

"Handbook of Physical Constants," J. Handin in S. B. Clark (ed.), sect. 2, p. 223, memoir 97, Geological Society of America, New York, 1966.

"Mechanical Twinning of Crystals," M. V. Klassen-Neklyudova, J. E. S. Bradley (trans.), Consultants Bureau, New York, 1964.

"Plasticity of Crystals," M. V. Klassen-Neklyudova, Consultants Bureau, New York, 1962.

"Plasticity of Crystals," E. Schmid and W. Boas, F. A. Hughes and Company, Ltd., London, 1950.

"Twinning," E. O. Hall, Butterworth & Co. (Publishers), Ltd., London, 1954.

SPECIFIC TOPICS

Andrade, E. N. da C.: Regular Surface Markings in Stretched Wires of Soft Metals, *Phil. Mag.*, **27**: 869 (1914).

—— and R. Roscoe: Glide in Metal Crystals, *Proc. Phys. Soc.*, **49**: 166 (1937).

Bilby, B. A.: *Progr. Solid Mechanics*, **1**: 331 (1960).

——, A. H. Cottrell, and K. H. Swinden: The Spread of Plastic Yield from a Notch, *Proc. Roy. Soc. (London)*, **272A**: 304 (1963).

—— and L. R. T. Gardner: *Proc. Roy. Soc. (London)*, **247A**: 92 (1958).

—— and E. Smith: *Acta Met.*, **4**: 379 (1956).

Borg, I., and J. Handin: Torsion of Calcite Single Crystals, *J. Geophys. Res.*, **72**: 641 (1967).

Buerger, M. J.: Translation Gliding in Crystals, *Am. Mineralogist*, **15**: 1, 21, 35 (1930).

Cahn, R. W.: Twinned Crystals, *Advan. Phys.*, **3**: 202 (1954).

Carpenter, H. C. H., and C. F. Elam: The Production of Single Crystals of Aluminum and Their Tensile Properties, *Proc. Roy. Soc. (London)*, **100A**: 329 (1921).

Ewald, P., T. Poeschl, and L. Prandtl: "The Physics of Solids and Fluids," Blackie & Son, Ltd., Glasgow, 1936.

Ewing, J. A., and W. Rosenhain: Experiments in Micro-metallurgy: Effects of Strain, *Proc. Roy. Soc. (London)*, **65**: 85 (1899); also, The Crystalline Structure of Metals, *Phil. Trans. Roy. Soc. (London)*, **193A**: 353 (1900).

Feynman, R. P., R. B. Leighton, and M. Sands: "The Feynman Lectures on Physics," vol. II, Addison-Wesley Publishing Company, Inc., Reading, Mass., 1964.

Frank, F. C., and N. Thompson: On Deformation by Twinning, *Acta Met.*, **3**: 30 (1955).

Gilman, J. J.: *Acta Met.*, **3**: 277 (1955).

——: *Trans. AIME*, **200**: 621 (1954).

—— and T. A. Read: Bend Plane Phenomenon in the Deformation of Zinc Monocrystals, *Trans. AIME*, **197**: 49 (1953).

Haasen, P., and A. W. Lawson: *Z. Metallkunde*, **49**: 280 (1958).

Hanafee, J. E., and S. V. Radcliffe: *J. Appl. Phys.*, **38**: 4284 (1967).

Handin, J.: "Handbook of Physical Constants," memoir 97, p. 238, Geological Society of America, New York, 1966.

Jaswon, M. A., and D. B. Dove: The Crystallography of Deformation Twinning, *Acta Cryst.*, **13**: 232 (1960).

Jillson, D. C.: Quantitative Stress-Strain Studies on Zinc Single Crystals in Tension, *Trans. AIME*, **188**: 1129 (1950).

Johnsen, A.: Die Struktureigen schaften der Kristalle, *Fortschr. Mineral.*, **3**: 93 (1913).

Liebisch, T.: *Nachr. Akad. Wiss. Gottingen, II. Math. Physik. Kl.*, p. 435 (1887).

McClintock, F. A.: Ductile Fracture Instability in Shear, *J. Appl. Mech.*, **25**: 581 (1958).

Mark, H., M. Polanyi, and E. Schmid: Strain-hardening of Crystals, *Z. Phys.*, **12**: 58 (1922).

Marsh, D. M.: "Fracture of Solids," Drucker and Gilman (eds.), Interscience Publishers, New York, 1963.

Mügge, O.: Uber Translationen and verwandte Erscheinungen in Krystallen, *Neues Jarbuch Mineral., Geol. Paleon.*, **I**: 155 (1898).

Nye, J. F.: *Acta Met.*, **1**: 153 (1953).

Polanyi, M.: Deformation von Einkristallen, III, *Z. Krist.*, **61**: 49 (1925).

—— and E. Schmid: *Z. Phys.*, **16**: 336 (1923).

Reusch, E.: Ueber eine besondere Gattung von Durchgängem im Steinsalz und Kalkspath, *Poggendorf's Analen*, **132**: 441 (1867).

Robinson, P. M., and H. G. Scott: Plastic Deformation of Anthracene Single Crystals, *Acta Met.*, **15**: 1581 (1967).

Rose, G.: *Abhandl. Berliner Akad.*, taf. 2, p. 57, Berlin (1869).

Schmid, E., and W. Boas: "Kristallplastizitaet" (1935), English ed., "Plasticity of Crystals," F. A. Hughes and Company, Ltd., London, 1950.

Starr, A. T.: Slip in a Crystal and Rupture in a Solid Due to Shear, *Proc. Phil. Soc. (Cambridge)*, **24**: 489 (1928).

Taylor, G. I.: Resistance to Shear in Metal Crystals, *Trans. Faraday Soc.*, **25**: 121 (1928).

—— and C. Elam: The Distortion of an Aluminum Crystal during a Tensile Test, *Proc. Roy. Soc. (London)*, **102**: 643 (1923); **108**: 28 (1925).

Von Goeler and G. Sachs: *Z. Phys.*, **41**: 103 (1927).

Von Mises, R.: Mechanik der Plastichen Formanderung von Kristallen, *Z. Angew. Math. Mech.* **8**: 161 (1928).

West, C. D.: *U.S. ONR Technical Report*, contract no. N7 ONR-39102 (1955).

Whapham, A. D., and H. Wilman: *Proc. Roy. Soc. (London)*, **237A**: 513 (1956).

4

DISLOCATION GEOMETRY

It is not the purpose here to present a formal treatment of the theory of dislocations. Only enough theory will be presented to provide a background for our discussion of plastic flow. Furthermore, the discussion will be as informal as possible, because this seems most appropriate for the present state of the subject. This has the advantage of tending to leave the theory flexible, so that any of a number of directions can be chosen for its future development. There recently have been several excellent and more formal discussions. Among the most recent have been those by Friedel (1964), Hirth and Loethe (1968), and Nabarro (1967).

The framework of dislocation theory is the geometry displayed by dislocation lines. It forms a basis for describing their properties. The great flexibility of these lines allows an exceedingly large variety of configurations to appear, but relatively few principles are involved. Later, discussions of energetics, forces, and motions will be supported by the geometrical framework that is first developed here.

4.1 CONCEPT OF A DISLOCATION LINE

The geometrical situation that we shall be particularly concerned with is shown schematically in Fig. 4.1, where the top half of a crystal is progressively displaced relative to the bottom half. It may be noted that, because of its periodicity, the internal structure (indicated schematically by a set of evenly spaced parallel lines) becomes restored after the displacement has occurred if the displacement vector has the proper magnitude. The boundary between the displaced and undisplaced regions is the *dislocation line*, and it is given this name because the amount of relative displacement changes discontinuously at it.

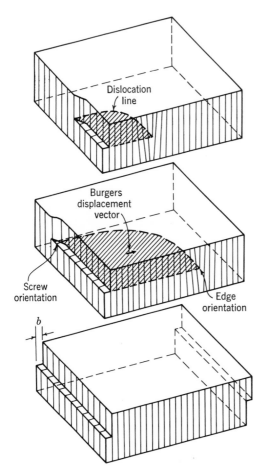

Fig. 4.1 Spread of translation-gliding across a crystal.

In a rather *general* sense, a dislocation line may be considered to be the boundary between two domains in space. Within each domain the displacement relative to a reference surface is different. At the boundary between the two domains, the displacement field changes discontinuously as indicated schematically in Fig. 4.2. In this figure, displacements exist above and below the reference surface inside the shaded domain. They are given by displacement vectors of the type $\mathbf{u}_r$. These vectors are defined in turn by position vectors in the coordinate system and are designated $\mathbf{r}_a$ and $\mathbf{r}_b$. Outside the shaded domain there are no displacements, so there must exist a boundary between the two domains somewhere on the reference surface S. It is this boundary which is called a *dislocation line*.

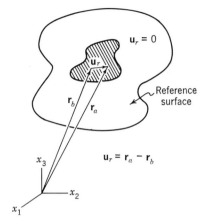

Fig. 4.2 Two domains in space within
which the displacement relative to a
reference surface is different.

The world abounds with examples of this kind of geometry, although a
formal description is not often applied. An immediate example occurs if one
picks up a piece of writing paper and proceeds to tear it into two pieces.
When the tear has progressed partway through the paper, the average displacement ahead of the tear is zero, whereas the displacements behind the
tear are large along the line of the tear, and there is a discontinuous change
in displacement at the tip of the tear. In this case the length of the dislocation line is very short (being equal to the thickness of the paper), but it can
be made longer by tearing a stack of paper sheets. There are, of course,
many other types of tearing or fracturing, all of which involve the passage
of one or more dislocation lines through a piece of material.

Another example can be found on a much larger scale in geological forma-
tions. These are the shear faults in the crust of the earth which give rise to
large and small earthquakes. Sometimes shear faults occur on a very small
scale, as in the sliding of one part of a sandpile relative to the rest during a
rain storm. Other cases are on a truly gigantic scale, as in the case of the
severe San Andreas Rift in California, an aerial photograph of which is
shown in Fig. 4.3. Here, the relative displacement of one side of the shear
plane relative to the other can be as much as 300 miles. (This is why the
structures on the two sides of the shear plane in the photograph do not
match at all.) In this case, the dislocation lines are at the ends of the fault
if the shear displacement field is uniform over the plane of shear. For further
information about the role played by these shear faults in earthquake phenomena, the interested reader is referred to an article by Benioff (1964).

Fig. 4.3   Aerial view of San Andreas Rift looking northwest over India Hills, California. (*Spence Air Photos, Los Angeles.*)

Dislocations may also occur on a grand scale during the motion of glaciers, because the high end of a glacier moves first relative to the low end, so the displacement field of the ice relative to the underlying earth is not uniform and bounding lines or dislocations must exist at some places along the interface between the ice and the earth.

Another natural science which contains many examples of dislocations is biology. This is especially clear during some animal locomotion processes. A millipede, for example, does not move its numerous legs simultaneously, but rather begins to move by lifting the legs near one end of its body and putting them down at a new position. Then, the next set of legs is lifted and replaced, so the end of the body is displaced while the rest of it retains the initial position. This creates a discontinuity in displacement somewhere along the length of the body, and this defines a dislocation line. As the dislocation propagates, the millipede moves. Similarly, earthworms and snakes move by partial displacements of their bodies either parallel or transverse to the lengths. As the dislocation between the displaced part of the body and the remainder moves, so does the snake or earthworm. Still other examples of dislocations in biological systems appear in the microscopic mechanism of muscular contraction and in the motions involved in the synthesis of proteins.

Many well-known mechanical devices contain dislocations and use them to accomplish some purpose. A common example is the zipper fastener. Here, the slide which divides the fastened portion of the system from the unfastened portion lies at the dislocation. Another example is the method that one uses to move a large rug across the floor of a room. Because of the large amount of friction between the rug and the floor, it is often convenient to lift only one edge of the rug and displace it slightly so as to produce a ruck. It is then relatively easy to push this ruck across the room, displacing a small amount of rug with each movement of the ruck until it emerges from the far end, and a displacement of the entire rug has been accomplished. This method requires only a small amount of force to be applied at any one time whereas it would require a very large force to slide the entire rug simultaneously across a rough floor.

It may be seen from these several examples that a dislocation is essentially a geometric configuration. Therefore, it can exist in the imagination, as well as in real physical systems, so certain non-Riemannian dislocated spaces constitute a topic in pure differential geometry.

In the discussion that follows, we are interested primarily in a special class of dislocations, those that exist in solid elastic bodies. Such solids can be continuous and contain continuous strains (i.e., displacement gradients) and yet contain multivalued displacements. It is these latter (illustrated by Fig. 4.4) which constitute the description of a dislocation in an elastic solid.

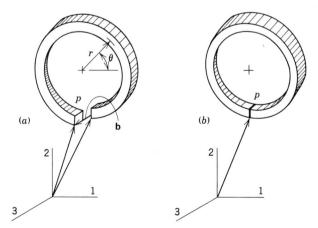

Fig. 4.4 Helicoidal ring at (a) has its ends closed together and welded at (b).

In the figure at (a) is shown a segment of a helical ring. One can imagine that this segment has been cut from a steel spring. The segment can be made continuous by performing the operation indicated by the vector **b** and then welding the two ends of the segment together as indicated at (b) in the figure. The ring at (b) is continuous and contains continuous strains, but its displacement field is multivalued. This may be seen as follows: Since the ring has a radius r, its circumference is $2\pi r$, so the strain in it which is defined as the displacement **b** divided by the height (or $2\pi r$) is equal to (letting **b** lie parallel to x_3)

$$\epsilon_{\theta 3} = \frac{|b|}{2\pi r} \qquad \text{``engineering strain''}$$

but, by definition,

$$\epsilon_{\theta 3} = \frac{1}{r} \frac{\partial u_3}{\partial \theta}$$

so

$$du_3 = \frac{b}{2\pi} d\theta$$

and, upon integration (since $u_3 = 0$ when $\theta = 0$),

$$u_3 = \frac{b}{2\pi} \theta$$

This relation states that u_3 increases indefinitely with θ (therefore, it is multivalued at any point), even though the strain $\epsilon_{\theta 3}$ is continuous and has a single fixed value. The fact that u_3 is multivalued may be seen by noting that the displacements at point p of Fig. 4.4b are both 0 and **u**.

A ring such as that of Fig. 4.4b is in a state of "self-stress," and such situations for elastic solids were first studied in detail by the school of Volterra (1907). They form a well-defined branch of the theory of elasticity.

The class of dislocations that interests us primarily is further restricted by the fact that we shall consider mainly crystalline solids. Crystals have discrete structures which are built up from atomic or molecular particles, so the displacements associated with dislocations in them are not only multivalued, but also quantized; that is, only definite translations or rotations are allowed. These are the ones which preserve the symmetry of the crystal and therefore allow continuous strains. Translational dislocations usually have much lower energies than rotational ones, so they are most commonly observed, and we shall concentrate on them. Figure 4.1 illustrates a simple translational dislocation. Here the crystal is sheared progressively over part of its cross section by a definite amount equal to b, a translation vector of the *crystal lattice* (then the dislocation is said to be a "perfect" one) or a translation vector of the *crystal structure* (said to be "imperfect" if this is not also a lattice vector). The center of the dislocation lies at the boundary between the part of the schematic crystal that has been sheared and the part that has not.

The strain that is associated with a translational dislocation can be produced in many ways, since the only essential feature is that the displacement be multivalued at any point. For example, in Fig. 4.5 several ways of making

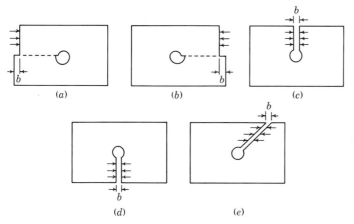

(a)   (b)   (c)

(d)   (e)

Fig. 4.5   Various methods for producing the same elastic translational-dislocation configuration.

the same translational dislocation are indicated schematically. It may be noted in this figure that methods (a) and (b) of producing the dislocation do not change the total mass of the body, whereas methods (c), (d), and (e) do require the mass to be changed.

In addition to its displacement vector, a dislocation line requires that a tangent vector lie parallel to it for a complete definition. Depending on the relative orientations of these two vectors, various limiting types of dislocations can arise that will be discussed in Sec. 4.4.

4.2 PHYSICAL UTILITY OF DISLOCATIONS

Dislocations are of fundamental importance in mechanics because they reduce the force needed to cause motion in a system. Like a lever, they allow a given amount of work to be done by a small force moved for a large distance, instead of a large force moving for a short distance.

The motion of dislocations in crystalline solids causes plastic flow, and the flow rate depends on the amount of displacement carried by each dislocation line, the concentration of dislocation lines, and the mean dislocation velocity. Quantitatively, the flow rate is given by the product of these three quantities.

It is interesting, and perhaps enlightening, to compare the plastic flow process with that of electrical conduction, as the two processes have several points of similarity, although they are not completely analogous. Whereas the flow of current in a solid consists of the motion of electrons in it, the flow of displacement (that is, plastic flow) consists of the motion of dislocations. Both processes involve a certain number of elements (electrons, dislocations) carrying a discrete quantity of something (charge, or displacement) at a certain velocity.

Both the electrons and dislocations can have either positive or negative character. They are driven by force fields (electric or stress) and depending on how fast they move under a given driving force, they are said to have a certain *mobility*. In both cases the solid in which they move presents a *resistance* to their motion and thereby tends to dissipate their kinetic energy as heat. Both entities can multiply as they move, thereby causing catastrophes such as dielectric breakdown or plastic yielding. Both entities have limits on their top speeds. In the case of electrons, this is the velocity of light; in the case of dislocations, it is the velocity of sound. It is also interesting to note that in both electrical conduction and plastic flow, the "action" per entity is large, while the effective mass is low. Therefore, the mobility is high, and flow is rapid even for small driving fields (electric or stress). Finally it may be noted that just as in electrical conduction, where the structure of a crystal is restored to its initial state after the passage of an electron, so the structure is restored after the passage of a dislocation has caused some plastic flow.

4.3 THE BURGERS VECTOR AND ITS CONSERVATION

A complete description of a dislocation line in a crystal is necessarily rather complicated. The line has a certain direction; there is a certain discontinuity in displacement at it; and, in some structures, the exact position of the center of the dislocation relative to the atomic structure may have a significant influence on its properties. Also, depending on the details of the atomic structure, the state of a dislocation may depend strongly on the direction of its line and on the direction of the shear displacement relative to the structure that occurs across it. Nevertheless, much of the character of a dislocation depends predominantly on the direction and magnitude of the displacement discontinuity that occurs at it. These are defined by the *Burgers vector*, whose significance may be understood by considering Fig. 4.6. Here,

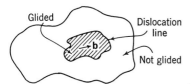

Fig. 4.6 The Burgers vector **b** specifies the direction and amount of displacement within a glided area (the displacement is constant over the glided surface).

a glided region is shown in a crystal, surrounded by a region that has not glided. A dislocation line separates the two regions, and the Burgers vector **b** specifies the direction and amount of displacement within the glided area. Note that the displacement is constant over the glided surface.

It is convenient to represent the Burgers vector **b** by means of Miller indices. If this is done, the orientation of a given Burgers vector, relative to the crystal structure, can be quickly recognized. Also, reactions between one or more dislocation can be written in a compact and relatively simple manner. The direction of the vector is given by means of the direction indices $[uvw]$. The magnitude is indicated by writing the lattice parameter d divided by a normalization factor N, in front of the bracket. The quotient d/N is chosen to be the smallest component of the Burgers vector relative to the crystal axes $\mathbf{a}_i$. Then the Burgers vector is represented as follows:

$$\mathbf{b} = \frac{d}{N}[uvw]$$

and the absolute magnitude of the Burgers vector is given by

$$|b| = \frac{d}{N}\sqrt{u^2 + v^2 + w^2}$$

This is illustrated in Fig. 4.7, which shows three Burgers vectors that may

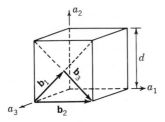

Fig. 4.7   Examples of Burgers
vectors in the cubic system.

be written as follows:

$$\mathbf{b}_1 = \frac{d}{2}[110]$$

$$\mathbf{b}_2 = d[100]$$

$$\mathbf{b}_3 = \frac{d}{2}[1\bar{1}0]$$

and a reaction between the three Burgers vectors may be written:

$$\mathbf{b}_2 = \mathbf{b}_1 + \mathbf{b}_3 = \frac{d}{2}[110] + \frac{d}{2}[1\bar{1}0] = d[100]$$

Finally, the absolute magnitudes of the three indicated vectors are

$$|\mathbf{b}_1| = \frac{d}{2}\sqrt{1^2 + 1^2} = \frac{d}{\sqrt{2}}$$

$$|\mathbf{b}_2| = d$$

$$|\mathbf{b}_3| = \frac{d}{\sqrt{2}}$$

The definition of the Burgers vector given above implies that it is related to a particular process, namely, the gliding process. Since the structure is the same on both sides of a dislocation, in reality there is no way of telling how an already existing dislocation has been formed. Therefore, it is useful to have a definition that does not depend on the process used to form the dislocation. Such a definition can be made in terms of the *Burgers circuit*. This is done by adding up the displacements along a line circuit that completely surrounds the dislocation and closes on itself. The sum of the displacements around this circuit is then compared with the sum of the displacements around a circuit of the same length that does not enclose the dislocation line; the difference between the two sets of displacements defines the

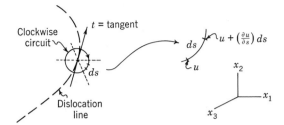

Fig. 4.8 Illustration of a Burgers circuit in a region where the displacement field is given.

Burgers vector. If the displacement-vector field is given, then Fig. 4.8 indicates how the circuit is made. And if one lets

$$\mathbf{u} = u_i = (u_1, u_2, u_3) = f(x_1, x_2, x_3)$$

be the displacement-vector field, then the circuit closure, or Burgers vector, is given by

$$\mathbf{b} = \text{circuit closure} = \oint \left(\frac{\partial u}{\partial s} \right) ds$$

If, instead, the *local crystal structure* is given, then a circuit can be formed by mapping the lattice points in the dislocated crystal back onto the corresponding perfect reference lattice (neglecting small distortions in the dislocated lattice). Any failure in complete closure of the circuit in the reference lattice is defined to be the Burgers vector. This definition relative to a reference lattice has the advantage that the Burgers vector then has a precise mathematical meaning and can be conveniently expressed analytically. This cannot be done directly in the dislocated lattice because this lattice is not well-defined from a mathematical viewpoint.

The above process can be seen clearly by studying Fig. 4.9, which shows a circuit AB drawn around a dislocation in the dislocated lattice at the left. The circuit consists of nineteen small steps, each of which is numbered. When these nineteen steps are mapped onto the perfect reference lattice at the right in the figure, it may be seen that after the nineteenth step has been taken, an additional step is required to get back to the initial position (a) from the nineteenth position (b). The magnitude of this step and its direction define the Burgers vector $\mathbf{b}$.

In order to preserve the usual definitions of mechanics and the early conventions of discussions of dislocations, it is important to choose a particular sense (direction) of the Burgers vector. The convention indicated in

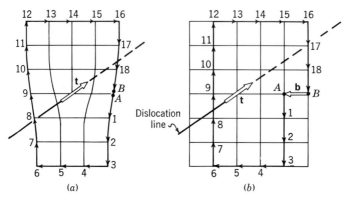

Fig. 4.9 Mapping of lattice points in a dislocated crystal (a) back onto the corresponding perfect reference lattice (b), neglecting small distortions in the dislocated lattice.

Fig. 4.9, originally set forth by Bilby, is the most compatible one. First, one chooses a tangent vector; then, if a clockwise circuit is made around the dislocation line such that a right-handed screw would advance in the direction of the tangent vector (as indicated in the figure), the Burgers vector of a positive edge dislocation (indicated at the left in the figure) points from right to left in the reference lattice and, hence, has a negative sense relative to the conventional choice of coordinate systems. However, as may be seen in Fig. 4.10, if a closed circuit is chosen in the reference lattice such that it

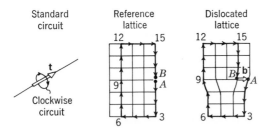

Fig. 4.10 Indicating that the sense of the Burgers vector in the reference lattice is opposite to that in the dislocated lattice. Note that the Burgers vector in the latter is positive when the extra half plane lies above the glide plane.

goes from A to B, then upon mapping the circuit into the dislocated lattice, as indicated on the right side of the figure, the resulting closure vector in the dislocated lattice goes from left to right, or in a positive sense, relative to a conventional coordinate system. Thus one obtains a positive edge dislocation in the dislocated lattice using the negative displacement in the reference lattice, as indicated in the previous figure. In this way, the conventions of mechanics are retained, as indicated by the sketches in Fig. 4.11; that is, the usual choice of the sense of a positive shear stress is retained,

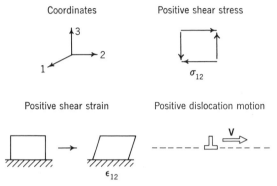

Fig. 4.11   Mechanical conventions that are preserved by the definitions of the text.

as well as that of a positive shear strain, and a positive shear stress tends to make conventional positive edge dislocations move in a positive direction relative to conventional coordinate axes.

Next we shall consider some conservation properties of Burgers vectors. These may be stated as follows:

1. Along a dislocation line the Burgers vector is uniform.
2. The sum of the Burgers vectors at a node between two or more dislocations is equal to zero (when the tangent vectors point outward from the node).
3. A dislocation line cannot end inside of a crystal.

The proofs of the three theorems above are quite simple. In the first instance, consider Fig. 4.12, showing a dislocation loop PQ with an associated Burgers vector $\mathbf{b}_1$. Suppose that the loop expands as indicated schematically at the right. It will then have the configuration $PQRS$, where the original configuration is indicated by the dashed line PQ. Suppose further that in moving from PQ the segment RS has acquired a Burgers vector $\mathbf{b}_2$ different from $\mathbf{b}_1$. Then a displacement difference must remain across PQ. But this was not given. Hence, $\mathbf{b}_1$ must be equal to $\mathbf{b}_2$, and the Burgers vector is the same everywhere along the given dislocation loop.

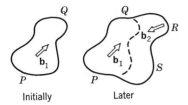

Fig. 4.12 $PQRS$ is a dislocation line, where RS has advanced from position PQ.

In order to prove theorem 2 above, we consider three dislocation lines meeting at a point, as indicated in Fig. 4.13. The directions of these three lines are indicated by the tangent vectors $\mathbf{t}_1$, $\mathbf{t}_2$, and $\mathbf{t}_3$ drawn pointing outward from the nodal point. Now, if standard Burgers circuits are constructed

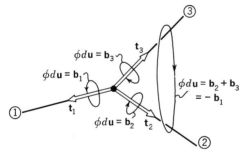

Fig. 4.13 Three dislocation lines that meet in space at a node and their associated Burgers circuits.

around each dislocation line, then the Burgers vectors $\mathbf{b}_1$, $\mathbf{b}_2$, and $\mathbf{b}_3$ are defined by the associated closure failures. But if a larger circuit is drawn that encompasses two lines (2 and 3 for example), then the closure failure of this large circuit must be equal to $-\mathbf{b}_1$, since the circuit is drawn in the same type of space as the one around line 1 but with an opposite sense. But the closure failure of the large circuit must also be equal to the sum of the Burgers vectors of lines 2 and 3; therefore, we have

$$\oint d\mathbf{u} = \mathbf{b}_2 + \mathbf{b}_3 = -\mathbf{b}_1$$

and so, we have proved that

$$\sum_{j=1}^{n} b_j = 0$$

Finally, we wish to prove theorem 3 above. Since a dislocation line is defined as the boundary of a *surface* (across which a discontinuity occurs), it must (1) close on itself, (2) end on an external surface, or (3) end at a node. It cannot end at a point, because then the surface that defines it would not have a complete perimeter.

4.4 THE SYMMETRY OF DISLOCATION LINES

STRAIGHT LINES

In general, the configuration of a dislocation line is a rather asymmetric pattern. However, in particular situations the configuration becomes much more simple and possesses certain elements of symmetry. The determining factor is the relative orientation between the Burgers and the tangent vectors. In general, an arbitrary angle can exist between them, allowing a multitude of possibilities, but only a few limiting types will be discussed here. Some of these are shown in Fig. 4.14, where the line tangent lies along

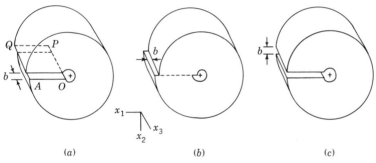

(a)   (b)   (c)

Fig. 4.14   Three basic dislocation types determined by the relative orientations of the tangent and the displacement vectors.

the x_3 axis in all cases, and the limiting cases are created by placing the displacement (Burgers) vector along the x_1, x_2, or x_3 axis.

The most symmetric case arises when the displacement and tangent vectors are parallel (Fig. 4.14a). Then, since the plane $AOPQ$ can be any plane containing the x_3 axis, the configuration has cylindrical symmetry in an isotropic elastic medium. In a crystal, the symmetry is reduced to that of the internal structure, as indicated schematically in Fig. 4.15 for the simple cubic case. Because of the fact that this dislocation causes the planes of the crystal that lie normal to the tangent vector to become segments of a helicoidal surface, it is called a *screw dislocation* (alternatively it is called

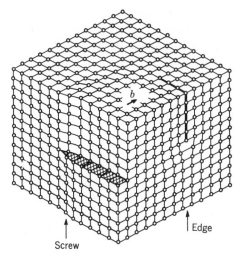

Fig. 4.15 Distortions of a crystal near the screw and edge orientations of a dislocation line.

the *Burgers dislocation*). Depending on the type of helical surface that it produces, the dislocation is said to be either right- or left-handed.

Because of its symmetry, all directions around a screw dislocation look the same in an isotropic medium. Therefore, it is not restricted to gliding motions along a single plane. This gives it comparative freedom of motion and has an important bearing on the overall plastic behavior of crystals that will be discussed in detail later. In crystals that are highly anisotropic, screw dislocations will, of course, prefer to move along the more weakly bonded planes; but they can move onto other planes if circumstances favor it.

If, as in Fig. 4.14b, the displacement vector lies parallel to the x_1 axis, and therefore is perpendicular to the dislocation line, a configuration results that is called the *Taylor*, or *edge, dislocation*. Depending on the sign of the shear distortion sketched in the figure, this type of dislocation is said to have either a positive or a negative character. Its symmetry is less than that of a screw dislocation. This can best be seen with the aid of Fig. 4.15, which shows how a shear displacement causes a "virtual relocation" of an extra half plane of atoms from the external surface of a crystal to the interior. The lower edge of this half plane lies at the dislocation line, and this accounts for the name given to this orientation. The presence of the extra half plane makes an edge dislocation antisymmetric about its own glide plane, whereas it has mirror symmetry about the half plane.

A final case arises if the displacement vector lies parallel to the x_2 axis and perpendicular to the tangent vector, as in Fig. 4.14c. If the displacement causes a gap to be opened up as sketched in the figure, then such a dislocation constitutes an elementary crack in the solid. If, on the other hand, the displacement is such as to close the indicated gap, then by comparison with Fig. 4.14b, it may be seen that the result is equivalent to the Taylor edge dislocation.

LOOPS

Dislocation lines commonly curve back on themselves to form loops. One type is called a *glide loop*, because it is formed when translation-gliding occurs locally inside of a crystal, as shown schematically in Fig. 4.16. Along

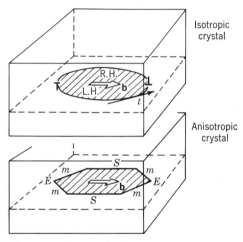

Fig. 4.16 Schematic dislocation loops of the glide type.

short segments of such a loop, the tangent vectors lie either parallel or perpendicular to the displacement vector, so the segments are of the pure screw or pure edge type, respectively. For the loop that is shown in the figure, the senses of the edge and screw components are given as they are conventionally defined. For an edge dislocation, this means that the extra half plane lies above the glide plane when the sense is positive. For a screw dislocation, if a *clockwise* circuit is made around the line (looking in the direction that the tangent vector points) and this causes an *advance* along the tangent vector (as a right-handed wood screw behaves), then the sense of the dislocation is right-handed.

Because the displacement and tangent vectors are perpendicular for edge dislocations, it is possible to form a complete loop consisting of only this dislocation type. (This is not possible for the screw type.) Such a loop is illustrated in Fig. 4.17 and is called a *prismatic dislocation* because it moves

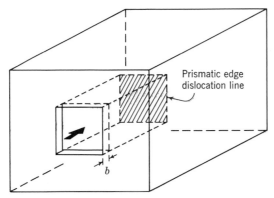

Fig. 4.17   Edge dislocation that glides along the faces of a square prism.

on the surface of a prism. In an isotropic medium the prism is circular; but it may also be elliptical, hexagonal, square, trigonal, or rectangular, depending on the symmetry of the crystal about its axis.

The edge and screw orientations of dislocation lines are not the only high-symmetry possibilities, because some segments of a loop that lies on a plane whose symmetry is trigonal or hexagonal may prefer to lie at 60 or 120° with respect to the Burgers vector (see segments marked m in Fig. 4.16). The most common shape for a loop will, however, be elliptical, so that small segments of it are of the edge and screw type, whereas most of it has a mixed character.

4.5 LOCALIZED LINE IRREGULARITIES

Since dislocation lines are basically geometrical configurations, they are quite flexible (exactly how flexible will be discussed later) and so can follow complex curves in three-dimensional space. Therefore, in addition to the straight-line configuration, they may follow simple curves, as indicated in Fig. 4.18a. Then, associated with each point along the line is a radius of curvature, as indicated in the figure. Also, they may follow double curves, and a particularly simple case is sketched in Fig. 4.18b. This shows a helical

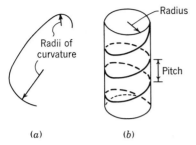

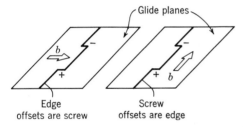

Fig. 4.18 Curved dislocation lines: (a) simple; (b) double.

dislocation line which is defined by a helical radius plus the pitch of the helix. Other more complex double or triple curves may also be followed.

The curvature of the core of a dislocation line may be very sharp with a radius as small as one atomic radius. This allows sharp bends to exist in the lines, which are known as kinks and jogs. Kinks are illustrated in Fig. 4.19.

Fig. 4.19 Kinks in dislocation lines.

They are associated with local offsets lying in the glide plane of the dislocation. Since the short segments of the offsets have different orientations from the main lines, their symmetry characteristics are different. In the case of a kink in an edge dislocation, as indicated at the left in the figure, the short segments lie parallel to the Burgers vector and therefore have screw character; whereas in the case of kinks in screw dislocations, the short segments lie perpendicular to the Burgers vector and therefore have edge character, as indicated at the right in the figure. Also, it should be noted that the kinks may have either a positive or a negative sense. Therefore, a pair of positive and negative kinks may annihilate by moving together, and a pair having the same sense can coalesce to form a superkink by moving together.

The name jog is associated with an offset in a dislocation which has a component that is perpendicular to the glide plane, as shown in Fig. 4.20.

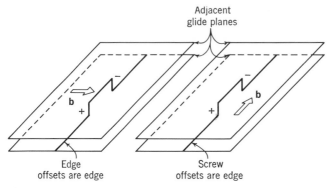

Fig. 4.20   Jogs in dislocation lines.

Here, the short offset segments have the same character as the main edge dislocation (left side of figure), since they are oriented perpendicular to the Burgers vector. On the other hand, for the screw dislocation (right side of figure), the offset segments are perpendicular to the Burgers vector and therefore have edge character. Thus, it may be noted that in an isotropic crystal, kinks and jogs in screw dislocations are equivalent (compare Figs. 4.19 and 4.20). Just as for kinks, jogs can have either a positive or negative sense and, hence, can annihilate one another or coalesce to form superjogs.

Since a jog does not necessarily lie on a plane that is crystallographically equivalent to the glide plane of the main dislocation, its mobility may be significantly less. In fact, in the case of an edge jog on a screw dislocation (Fig. 4.20), the mobility of the jog in the direction of motion of the main dislocation is zero. This is because the glide plane of the edge-dislocation segment lies parallel to the screw dislocation line, and as we have pointed out previously, an edge dislocation has zero mobility perpendicular to its glide plane. A jog with low mobility can cause a cusp to form in a moving dislocation line, as illustrated in Fig. 4.21. Such cusps play an important role in the regeneration of dislocations and in strain-hardening.

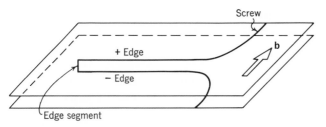

Fig. 4.21   Cusp in a moving screw dislocation line.

4.6 INTERSECTIONS OF DISLOCATIONS

As dislocation lines move about in three-dimensional space, they may sometimes approach each other in a parallel-line configuration and then pass by
and separate; or they may interact through their stress fields and become
bound together, as we shall discuss in more detail later. If two approaching
lines are not parallel to one another, then an intersection is likely to take
place. In general, there are a large number of possible intersection geometries,
and we shall only describe the most symmetric cases.

The geometry that arises when two edge dislocations intersect is indicated
in Fig. 4.22. Here the edge dislocation in the horizontal plane is taken to be

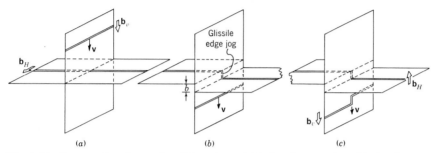

Fig. 4.22   Intersection of two edge dislocations: (a) before intersection; (b) after intersection (case of perpendicular Burgers vectors); (c) after intersection (case of parallel
Burgers vectors).

fixed, and when the second dislocation (moving in the vertical plane) passes
through the horizontal one, it causes an offset to occur. If the Burgers
vectors of the two intersecting edge dislocations are perpendicular, as indicated in Fig. 4.22b, the moving dislocation moves on after the intersection
without any change in its configuration, leaving an edge jog in the horizontal dislocation. Since this edge jog lies perpendicular to the Burgers
vector of its dislocation, it can move rapidly in a plane parallel to the Burgers
vector and is said to be *glissile*. On the other hand, if the Burgers vectors
of the two intersecting dislocations are parallel, as at c in the figure, then both
dislocations change their configuration during the intersection by acquiring
kinks. In the figure, the dislocations lie on perpendicular planes, but any
crystallographic inclination is possible.

The case of an edge dislocation intersecting a screw dislocation is illustrated by Fig. 4.23. The screw dislocation has converted the planes lying
normal to it into a helical ramp. Therefore, an edge dislocation that lies on
one level prior to the intersection acquires two levels during the intersection

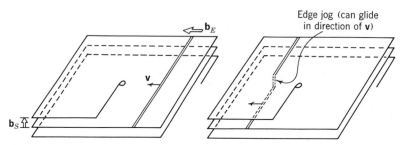

Fig. 4.23 Intersection of a moving edge dislocation with a stationary screw dislocation.

and thus contains an edge jog after the intersection. This jog lies in a plane parallel to the Burgers vector of the dislocation and therefore is of the glissile type. The same result is obtained if the screw is the moving one. It intersects an edge dislocation of the same form as that in Fig. 4.23. However, a moving screw dislocation might also intersect an edge dislocation having its Burgers vector parallel to that of the screw. Such an intersection would cause the edge dislocation to acquire a kink.

The type of dislocation intersection that disturbs the structure of a crystal most arises when two screw dislocations intersect. In this case, as may be seen in Fig. 4.24, the moving horizontal dislocation acquires a jog (and, by

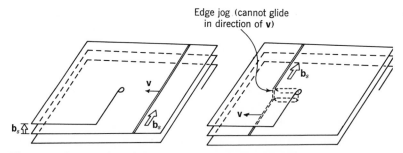

Fig. 4.24 Intersection of a moving screw dislocation with a stationary screw.

symmetry, so does the stationary vertical dislocation). The jog in horizontal dislocation is directed perpendicular to the Burgers vector of this dislocation, so it has edge character. Its glide plane lies parallel to the screw dislocation line, so it can readily move along the line, but it cannot move in the direction of the velocity vector (indicated in the left-hand sketch of the figure). Since it cannot move, such a jog causes a cusp to form in the moving screw dislocation. As we shall see shortly, such a cusp is equivalent to a row of vacancies or interstitial atoms in structure and therefore represents a very severe disturbance.

4.7 REGENERATION (MULTIPLICATION)

In passing completely across a crystal, an individual dislocation line produces a plastic shear strain of only about 10^{-8}. Therefore, numerous dislocations are required to produce the large strains that are commonly observed (sometimes as large as 5 or more). An essential part of the theory of plastic flow concerns itself with the mechanisms by which dislocations can be created during the flow process. Such mechanisms are needed not only to explain the large observed plastic strains, but also to explain the large increase in dislocation density that is observed to occur. We now have the background to discuss the geometrical aspects of various multiplication mechanisms (their energetics will be discussed later). There are several classes, and we shall discuss these in turn. They include nucleation of new dislocation loops under stress, dynamical mechanisms, quasi-static regeneration, and breeding mechanisms.

NUCLEATION OF NEW LOOPS

If a dislocation loop forms in a crystal that is stressed by external forces, the displacements associated with the loop allow the external forces to do work, thereby tending to reduce the potential energy of the system. Thus, a driving force exists, tending to cause a spontaneous formation of dislocation loops. However, the self-energy of a dislocation loop is relatively high, so for ordinary stress levels (that is, stresses of order 10^{-3} to $10^{-2}G$†) a dislocation loop in starting from zero size must expand to a diameter of about 10^{-4} cm before enough external work is done to compensate for the increase in energy associated with the self-energy of the dislocation loop. The result is that loops can be spontaneously nucleated only at very high stress levels, approaching $G/10$. Furthermore, thermal agitation is not very helpful because of the relatively large size that must be reached before an embryonic dislocation loop becomes stable (Cottrell, 1953, p. 53). The conclusion is that one cannot expect many loops to form spontaneously during ordinary plastic flow conditions.

In some crystals there may exist small notches or hard precipitate particles which cause concentrated stresses that are great enough for nuclea-

† G = shear modulus.

tion (Gilman, 1959), but this requires special circumstances, so it cannot be a general mechanism.

The fact that spontaneous dislocation nucleation is so difficult has caused people to search for other mechanisms. The most notable of these searches was led by Frank (1950), and he ultimately discovered how dislocations can regenerate themselves at quite low stress levels while moving at low velocities (Frank and Read, 1950). Before and after this discovery, a number of other regeneration mechanisms were proposed and these mechanisms will be described in outline here; the reader is referred to original sources for detailed discussions.

DYNAMICAL MECHANISMS

One way in which a dislocation might regenerate itself would be to store energy kinetically as it was collected from the applied stress field (Fig. 4.25). At the left in the figure, a dislocation moves from left to right. The configura-

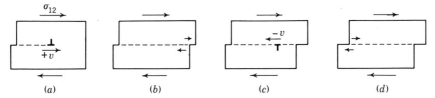

(a) (b) (c) (d)

Fig. 4.25 Frank's dynamical reflection mechanism for regeneration at a free surface: (a) moving positive edge dislocation; (b) dislocation gone, momentum remains; (c) reflected negative edge dislocation; (d) negative dislocation gone, momentum remains.

tion at the instant it reaches the right-hand surface is sketched in the second drawing. Here the dislocation has just disappeared by moving out through the free surface, but the momentum associated with its motion in the crystal remains. This momentum tends to make the upper half of the crystal overshoot, thereby tending to produce a negative edge dislocation that would move toward the left in the given applied stress field, as sketched in the third drawing. When the negative dislocation moves out of the left-hand free surface, its momentum tends to make the bottom half of the crystal overshoot, thereby producing a positive edge dislocation moving left to right. Thus, the cycle is completed. It may be noted that this mechanism does not increase the number of dislocations in a crystal, but only holds it constant.

In order for a dynamical mechanism to be feasible from the point of view of energetics, the kinetic energy of the dislocation at the time it reaches a free surface must be approximately equal to the static self-energy. This means that the velocity of the dislocation must be a substantial fraction of the velocity of sound in the crystal. However, for a single dislocation in an otherwise perfect crystal, the motion is too highly damped in most cases for the kinetic energy to reach a value approximating the self-energy. Therefore, this mechanism is not usually viable. However, after some prior strain, surface steps will exist at the ends of the glide plane; they will create shear-stress concentrations, making it easier for this mechanism to operate, so it may play a role during the later stages of deformation.

Another dynamical mechanism (originally proposed by Frank) involves internal crossing-over of dislocations, and it is illustrated by Fig. 4.26. A

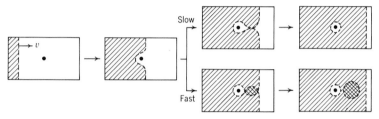

Fig. 4.26   Multiplication of a dislocation line via dynamic internal crossing-over.

dislocation moving from left to right with velocity v is shown as it moves across a glide plane. An obstacle lies on the glide plane, and in the second drawing of the figure, the dislocation has reached the obstacle and has begun to move around it. The remainder of the figure divides into two branches, one showing subsequent events for a slowly moving dislocation, and the other showing what happens if the dislocation is moving very fast. In the slow case, the dislocation simply wraps around the obstacle until segments of it having opposite character meet and annihilate. After this has occurred, a loop remains that encircles the obstacle (unless the obstacle shears, allowing the dislocation loop to collapse), and the original dislocation line moves on, as shown in the drawing at the far right. For a fast-moving dislocation, when the segments of opposite character meet after the dislocation has wrapped around the obstacle, the momentum of the colliding segments allows them to cross over one another, thereby creating a new loop as at the bottom far right. Note that unlike the previous case, this mechanism increases the number of dislocations. However, it is again an unlikely mechanism, because drag forces prevent the kinetic energies of the two colliding dislocations from becoming sufficiently large unless they are aided by things like stress concentrations at internal boundaries.

QUASI–STATIC MECHANISMS

In 1950, Frank and Read realized almost simultaneously that, because dislocations are flexible and can move in two dimensions with considerable freedom, it is possible for multiplication to occur by means of various folding processes. An analog of this in three dimensions is the regenerative formation of soap bubbles at the mouth of a toy pipe. Each time a bubble is blown off the pipe, the soap membrane that initially closes it is regenerated, so the process can repeat. The formation of a single-ended, spiral dislocation line is perhaps the most simple geometrical case of quasi-static multiplication. The initial situation is illustrated at the top in Fig. 4.27, which shows a dis-

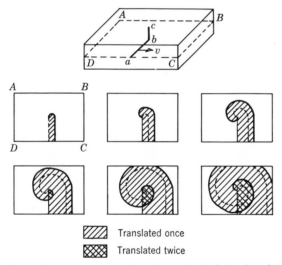

Translated once

Translated twice

Fig. 4.27 Single-ended, Frank-Read spiral developed by constant velocity motion of a line that is pinned at one end.

location segment ab lying in a plane $ABCD$ and moving with velocity v. One end a of this segment lies at the free surface of the crystal, and the other end b is at a node where it connects with a stationary dislocation segment bc that intersects the plane $ABCD$. The next drawing in the figure views the glide plane from above and shows the configuration that is generated when the mobile dislocation segment moves a small amount at constant velocity along its length. Subsequent drawings show how upon further motion a spiral configuration finally results. Since the length of the spiral is considerably greater than that of the initial segment, the total length of dislocation line per unit volume can be increased markedly through this process.

A closer analog with the soap-bubble case arises when a mobile dislocation segment ends on two nodes inside the crystal, as illustrated in Fig. 4.28a. Here two spirals begin to form (Fig. 4.28b, c, and d), but then the segments of these spirals which have opposite character meet and annihilate (e), so a loop finally results (f), and the initial mobile segment ab is restored.

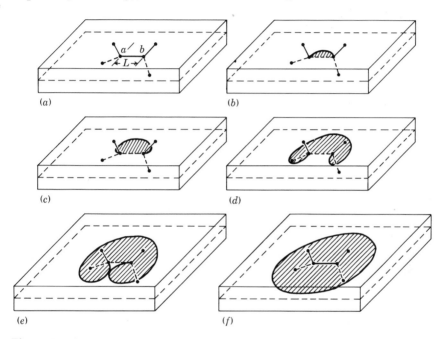

Fig. 4.28 Schematic operation of a double-ended, Frank-Read dislocation regenerator.

The radius of curvature of the dislocation line in this case can be seen to change as the process proceeds. The average curvature (that is, the reciprocal of the radius) increases as area is swept out until it reaches a maximum value equal to $2/L$, and then it decreases. Thus, a critical geometrical configuration arises during the process (when the mobile segment has become a semicircle), and associated with this is a critical stress, whose magnitude will be determined later.

For constant dislocation velocity and a constant rate of production of loops, this mechanism increases the dislocation length in proportion to the square of the time.

BREEDING DURING GLIDE

For the mechanisms discussed thus far, regeneration occurs at a fixed number of places in a crystal. Multiplication will occur more rapidly, however, if the number of regeneration sites increases during plastic deformation, that is, if breeding occurs, rather than just production. A few processes that can cause breeding will now be discussed.

The drawings in Fig. 4.29, for example, show how the intersection of a moving screw dislocation with several others that have displacement com-

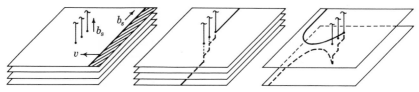

Fig. 4.29 Intersection of a moving screw dislocation with three stationary ones to form a superjog.

ponents normal to the glide plane can lead to the formation of a superjog. This retards the motion of part of the moving dislocation and produces the conditions needed for the Frank-Read mechanism to operate.

Another means for making a center of regeneration is a line-lengthening mechanism suggested by Orowan (1954) and illustrated in Fig. 4.30.

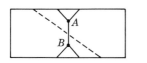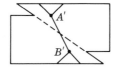

Fig. 4.30 Lengthening of a line segment through shear.

Here a dislocation segment AB, whose length is too short to allow it to operate as a Frank-Read source at the beginning, is lengthened during shear deformation to some length $A'B'$ that is sufficiently long.

Perhaps the most effective of all breeding mechanisms is the one called multiple cross glide (originally suggested by Koehler, 1952). At the left in Fig. 4.31 a screw dislocation moves along on a primary glide plane. If the motion of a segment of such a dislocation is somehow disturbed, the segment may move off the primary glide plane onto a cross glide plane, as sketched at the middle of the figure. Later, the motion of the segment will tend to

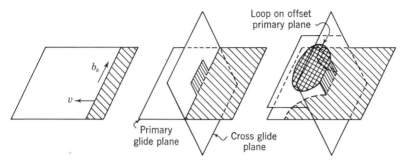

Fig. 4.31   Regeneration of dislocations via multiple cross glide.

revert to the primary glide plane, thereby producing two superjogs which hold back parts of the moving dislocation. If these jogs have sufficient height, a configuration is created that can act as a Frank-Read source, as at the right of the figure. Subsequently, a segment of the original dislocation line may be disturbed, so that further cross-gliding occurs. Therefore, the rate of breeding becomes proportional to the existing line length L, and if α is the breeding coefficient, then $\dot{L} = \alpha L$, so upon integration $L = L_0 e^{\alpha t}$. Because the line length increases exponentially with time, this process tends to dominate all others shortly after plastic deformation has begun.

Another deformation phenomenon that tends to cause dislocation breeding is the production of surface notches by shear offsets. These notches cause stress concentrations and, hence, favorable sites for dislocation nucleation or the operation of nearby sources. The notches may arise at external free surfaces or at internal grain boundaries or other interfaces. Since the stress-concentration factor is proportional to the square root of the size of an offset, such notches become increasingly effective as their offsets become larger. Eventually, they may even lead to strain-softening.

4.8 ARRAYS OF DISLOCATIONS

When the dislocation density within a material increases during plastic flow, the dislocations do not remain randomly distributed,† but take on certain special configurations which minimize their elastic energies. We shall consider the elastic interactions later, but shall now outline some of the configurations that can arise and are observed.

† The "tangles" observed with transmission electron microscopy are sometimes misleading because they are seen *post facto* and because it is very difficult to avoid heating of specimens before or during observation.

COMPACT GROUPINGS (MULTIPOLES)

Groupings of dislocation lines that lie parallel to one another are the most simple arrays and are often the most stable. Except in anisotropic crystals, parallel screw dislocations are not stable, however, because of their ability to glide on any plane. This allows them to mutually annihilate with ease. In contrast, edge dislocations that lie mutually parallel are able to form metastable arrays, because they cannot move off their own glide planes, and so they cannot mutually annihilate (unless they happen to lie on the same plane or the temperature is high enough for diffusion). A variety of the possibilities are sketched in Fig. 4.32. First, a *monopole* or single edge dis-

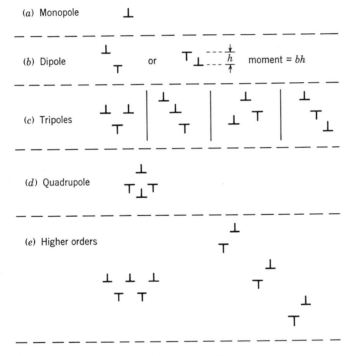

Fig. 4.32 Schematic edge-dislocation multipoles.

location is indicated. Next is a group of two edge dislocations of opposite sign, which is given the name *dipole*. Such a pair is characterized by the spacing of the edge dislocations that constitute it. Associated with this is a *dipole moment* equal to a Burgers displacement times the glide-plane spacing h. Notice that two distinctly different types of dipoles are possible, depending on whether the extra half planes of the edge dislocations point outward or inward. Furthermore, a pair of edge dislocations has two distinctly different possible metastable orientations, as sketched in the figure. Thus, a dipole is able to flip-flop from one orientation to another without being destroyed, and this provides a mechanism for internal energy absorption.

Under favorable circumstances an edge dipole can trap a third dislocation to form a metastable *tripole*, and four distinctly different arrangements are possible within a tripole (see Fig. 4.32c). Not all arrangements of three dislocations are stable. For example, three lines on a single glide plane make an unstable grouping.

A stable tripole may trap yet another dislocation to form a *quadrupole*. In this case, there are many possible arrangements of the four dislocations, and the stabilities of only a few of these have been investigated, but one that has stability is the symmetric one (Fig. 4.32d). When still more dislocations are arranged appropriately, various higher-order *multipoles* can be formed. A few of these are sketched in Fig. 4.32e, but the possibilities are by no means exhausted.

Since isolated parallel dislocation pairs are rare in real crystals, the above arrays most often arise when parts of loops interact, and different segments of the same loop may interact differently with neighboring loops. The result is a bewildering number of possible structures, and the only feasible description is a statistical one.

QUEUES

In the absence of an applied shear stress, more than one dislocation line lying on a single glide plane is an unstable group, and the dislocations will repel one another until they pass out of the crystal. However, if a stress is present, and somewhere along the glide plane there is an obstacle to motion, a group of dislocations may become queued up at the obstacle (Fig. 4.33).

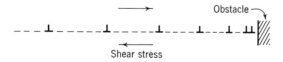

Fig. 4.33   Queue of edge dislocations.

Such an array strongly concentrates shear stress at its leading end. Therefore, it may lead to fracture or other processes that require a large amount of stress to get them started. The positions of the dislocations in a queue are related in a simple fashion that is the same as the electron orbits in a hydrogen atom; that is, if the leading dislocation is given an index 1 and it lies a unit distance away from the zero coordinate, then the second dislocation (index 2) lies a distance 2^2, or 4, from the reference zero. The third lies a distance 3^2, or 9, away, etc. For further details see Cottrell (1953).

TILT BOUNDARIES

In contrast to the parallel grouping of a dislocation queue, an array may lie perpendicular to the glide planes of a set of edge dislocations. This array produces a tilt between two sections of crystal as indicated in Fig. 4.34. The

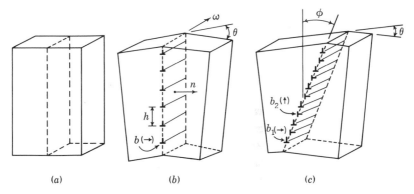

(a) (b) (c)

Fig. 4.34 Tilt boundaries formed by vertical arrays of edge dislocations: (a) monocrystal; (b) bicrystal with symmetric tilt boundary; (c) bicrystal with asymmetric tilt boundary.

half planes of the edge dislocations add their displacements to make a wedge of material. Insertion of this between the matrix material on either side causes a relative tilting of the crystal sections about the rotation axis ω. If there are n dislocations in the vertical array, then the sine of half the misorientation angle θ is given by

$$\sin \frac{\theta}{2} = \frac{nb}{2H}$$

and, for small values of θ,

$$\theta \simeq \frac{nb}{H}$$

If the edge dislocations are uniformly distributed with the spacing $h = H/n$, then the distance between the dislocations is related to θ and b:

$$h = \frac{b}{2 \sin \theta/2} \sim \frac{b}{\theta}$$

The simplest case is the symmetric one indicated in Fig. 4.34b. However, it is also possible for asymmetric boundaries to form in which the plane of the array is not perpendicular to the glide planes of the dislocations. This is sketched in Fig. 4.34c, and if the angle between the plane of the array and the perpendicular plane is called ϕ, then there will be two kinds of edge dislocations in the array, and their spacings will be given by the following relations:

$$h_1 = \frac{b}{\theta \cos \phi} \qquad \text{for } \theta \ll 1$$

$$h_2 = \frac{\theta}{b \sin \phi} \qquad \text{for } \theta \ll 1$$

The geometrical theory of tilt boundaries has been confirmed by comparing spacings of dislocations (as revealed by etch pits at their emergent points) and the tilt angles as obtained from x-ray measurements. Some of the results are listed in Table 4.1, and it may be seen that there is good agreement between theory and experiment.

TABLE 4.1 COMPARISON OF OBSERVED DISLOCATION SPACINGS WITH SPACINGS CALCULATED FROM MEASURED TILT ANGLES FOR GERMANIUM BICRYSTALS [after Vogel, Pfann, Corey, and Thomas, *Phys. Rev.*, 90:489 (1953)]

SPECIMEN	Θ_{meas} (sec)	h_{calc} (microns)	h_{obs} (microns)
A	17.5 ± 2.5	4.7 ± 0.7	5.3 ± 0.3
B	65.0 ± 2.5	1.3 ± 0.1	1.3 ± 0.1
C	85.0 ± 2.5	0.97 ± 0.2	0.99 ± 0.2

TWIST BOUNDARIES

As indicated in Fig. 4.35, crossed pairs of screw dislocations can cause a small rotation off the top of a crystal relative to the bottom if the dislocations lie in a horizontal plane. For a particular rotational angle θ, two sets of screw dislocations are required, and their spacing is given by the following

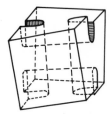

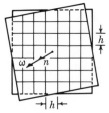

Fig. 4.35 Schematic twist boundaries formed by crossed grids of screw dislocations: (a) shows how the twist arises; (b) shows relation of grid spacings to rotation angle ω.

relation, as may be seen from the figure:

$$h = \frac{b}{2 \sin \theta/2} \simeq \frac{b}{\theta} \qquad \text{for } \theta \ll 1$$

GENERAL DISLOCATION BOUNDARY

The rotation in space of one piece of crystal relative to another has three degrees of freedom, which can be described in terms of three rotation angles about three spatial axes. In addition, the boundary plane has two degrees of freedom, and its orientation can be described in terms of two rotations about axes lying in the plane of the boundary. For the most general kind of dislocation boundary then, there are five degrees of freedom altogether. Frank has developed the theory of such boundaries, and Fig. 4.36 shows how a general boundary can be described in terms of the rotation axis ω about which crystallite II is rotated relative to crystallite I. The orientation of the boundary plane is described by the unit normal vector $\mathbf{n}$, and the rotation is $\theta\omega$. To find the closure failure of a Burgers circuit taken around a section of the boundary, one defines some vector $\mathbf{v}$ that lies in the boundary and has the Miller indices hkl relative to crystallite II. Then one also defines the vector $\mathbf{v}'$ which lies in the boundary and has indices $h'k'l'$ with respect

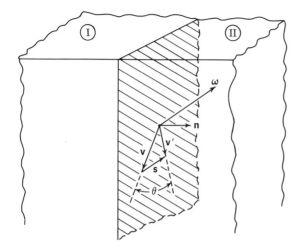

Fig. 4.36 Drawing of a generalized boundary between two crystallites. The boundary has five degrees of freedom. The vectors shown are: $\mathbf{n}$ = boundary normal; $\boldsymbol{\omega}$ = rotation axis for II relative to I; $\mathbf{v}$ = vector lying in the boundary—its indices are hkl relative to I; $\mathbf{v}'$ = vector lying in the boundary with indices $h'k'l'$ relative to II $(h = h',\ k = k',\ l = l')$; $\mathbf{s}$ = closure-failure vector.

to crystallite I. The two vectors are chosen to have the same positions relative to their associated crystallites, so their indices are equal (that is, $h = h'$, $k = k'$, $l = l'$), but relative to a single coordinate system the two vectors have different indices. The closure failure (given by a vector $\mathbf{s}$) is simply the vector difference of $\mathbf{v}$ and $\mathbf{v}'$, so the following relations hold:

$$\mathbf{s} = \sum_1^n b_i = \boldsymbol{\omega} \times \mathbf{v}$$

where the b_i are the displacements of the dislocations that are cut by $\mathbf{v}$. Thus, for small misorientation angles and a choice of $\mathbf{v}$ to make it a unit vector, the density of dislocations is simply $(\boldsymbol{\omega} \times \mathbf{v})\theta$.

For larger angles of misorientation, it is convenient to choose a median lattice as a reference. Then the two crystallites can be formed by equal but opposite rotations $|\omega_1| = \theta/2$ and $|\omega_2| = \theta/2$ about a single axis $\mathbf{L}$ that lies in the median lattice. The closure vector is given by

$$\mathbf{s} = (\boldsymbol{\omega} \times \mathbf{v})2 \sin \frac{\theta}{2}$$

Also, one can define an inclination vector

$$\mathbf{Z} = \mathbf{L}\left(2\sin\frac{\theta}{2}\right)$$

which has a tilt component

$$\mathbf{Z}_t = \mathbf{n} \times (\mathbf{Z} \times \mathbf{n}) = \mathbf{Z} - (\mathbf{Z} \cdot \mathbf{n})\mathbf{n}$$

and a normal twist component

$$\mathbf{Z}_n = (\mathbf{n} \cdot \mathbf{Z})\mathbf{n}$$

In terms of these tilt and twist components the closure vector can be written as follows:

$$\mathbf{S} = \mathbf{V} \times \mathbf{Z}_t + \mathbf{V} \times \mathbf{Z}_n$$

For a detailed discussion of boundary arrays in theory and experiment, the reader is referred to a review by Amelinckx and Dekeyser (1959).

NETWORKS

In addition to planar arrays of dislocations made up from sets of parallel dislocation lines, it is possible to form three-dimensional networks. These simply consist of segments of dislocation lines connected together at nodes. Typically these nodes are junctions of three or more dislocation lines that have arbitrary directions in space. This allows networks, reminiscent of the molecular structure of glass, to be built up in three dimensions. Using nodes at which four dislocation segments join, it is possible to develop a tetrahedral network like that of the structure of diamond. These are only the most regular possibilities, and there are of course many possible irregular structures, as well as more complex regular ones.

4.9 PLASTICALLY CURVED LATTICES

The phrase *plastically curved lattice* means that the net elastic stress of the configuration is zero, in contrast to the case of an elastically curved lattice. Also, a plastically curved lattice is different from a plastically curved shape, a distinction that is sometimes neglected. Figure 4.37 illustrates this distinction. On the left is shown a curved shape that has been generated by plastic deformation on a set of parallel glide planes. Such a curved shape does not necessarily contain dislocations, and the internal lattice is not curved. On the other hand, on the right a curved shape has been formed

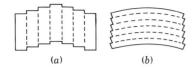

Fig. 4.37 Crystals with similar external curved shapes, but different internal structures: (a) lattice not curved—does not necessarily contain dislocations; (b) lattice curved—necessarily contains dislocations.

using a set of glide planes lying perpendicular to those of the case on the left. For this case, the material necessarily contains dislocations, and the internal lattice, as well as the external shape, is curved.

It should be noted that the discussion here considers only long-range lattice curvature. There can also be very short-range curvature near the centers of dislocations, and within edge-dislocation dipoles.

SINGLE CURVATURE AXIS

The most simple case to be considered arises when there is curvature about a single axis. This can be produced, as may be seen in Fig. 4.38, by a random array of edge dislocations which all have the same sign. If a Burgers circuit

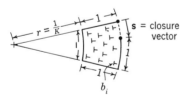

Fig. 4.38 Curvature of a crystal lattice caused by an excess of edge dislocations having the same signs.

is taken around such an array, the closure vector s is equal to the number of enclosed dislocations times the Burgers vector of each one (assuming they are all the same). If the area enclosed is a unit area, as indicated in the figure, this is the dislocation area density times the Burgers vector. The figure also shows that if the area enclosed by the Burgers circuit is taken to be a segment of a circular annulus, then the segment angle θ is equal to $1/r_0$. This in turn is given by

$$\frac{1}{r_0} = \frac{1 + \rho b}{1 + r_0}$$

and upon cross multiplying and canceling the unit length, one obtains for the curvature:

$$K = \rho b = \frac{1}{r_0}$$

Thus, the lattice curvature equals the dislocation density times the Burgers displacement. This has been experimentally confirmed for the case of plastically bent germanium crystals, as shown in Table 4.2.

TABLE 4.2 COMPARISON OF OBSERVED EXCESS EDGE DISLOCATION DENSITIES WITH THOSE EXPECTED FROM THE EXTERNAL CURVATURE OF BENT GERMANIUM CRYSTALS [after F. L. Vogel, Trans. AIME, **206:946 (1956)]**

BEND RADIUS (cm)	OBSERVED DISLOCATION DENSITY (10^6/cm^2)	CALCULATED DISLOCATION DENSITY (10^6/cm^2)
5	6.2	7.0
7	4.7	5.0
9	3.9	3.9
11	2.9	3.1
13	3.2	2.6
15	2.5	2.2
22.5	1.4	2.4

THREE CURVATURE AXES

The case of three-dimensional curvatures is more complicated than the one-dimensional case because it becomes possible for curvatures to arise through combinations of twisting and bending, rather than bending alone. Here the theory was originally developed by Nye (1953) and has since been generalized by Kondo; Bilby, Bullough, and Smith; as well as by Kroner. (See Bilby, 1960, for a review.)

It should be noted at the outset that we do not consider the curvatures that occur over very small distances, such as atomic dimensions, in this discussion. That is, if α is the distance between dislocations and l is some path that is long compared with this ($l \gg \alpha$), then l has the same length in both the curved and uncurved lattices. This condition can be extended by making α small, but as α decreases at constant curvature, the dislocation density ρ increases. This must be compensated by letting the Burgers displacements of the dislocations decrease. By letting the Burgers displacement tend toward zero as the dislocation density tends toward infinity, we can keep the curvature K constant. In this way a continuous distribution of infinitesimal dislocations can be discussed. Such a description for various geometrical situations is quite useful; but it should always be remembered that as soon as one puts the dislocations into motion, such a fictitious description loses its utility, because the most significant processes are then atomic interactions, and these are necessarily quantized.

We begin our detailed discussion by considering the problem of forming a Burgers circuit about an arbitrary vector that lies in a dislocated lattice. Such a vector is sketched in Fig. 4.39a, and a Burgers circuit which encloses

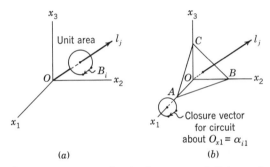

(a)   (b)

Fig. 4.39   Burgers circuits in a general dislocated lattice.

the unit area is also sketched where B_i is the closure vector of the circuit. The problem then is to relate the vector B_i to the arbitrary unit vector l_j. To do this, we let ABC be a unit triangle normal to l_j, as indicated in Fig. 4.39b. Then we form Burgers circuits (of unit area) around each of the coordinate axes Ox_1, Ox_2, and Ox_3. The closure vectors of these circuits are called α_{i1}, α_{i2}, and α_{i3}, respectively. The Burgers vector of a circuit around a larger area $OBCO$, for example, then is $\alpha_{i1}l_1$, where l_1 is the component of the vector l_j along the x_1 axis. Similarly, the Burgers vector of a circuit formed about $OCAO$ is given by $\alpha_{i2}l_2$, and for the case of $OABO$, $\alpha_{i3}l_3$.

As a result of the above formulations, the closure vector for the triangular circuit $ABCA$, which encloses and lies normal to the vector l_j, is

$$B_1 = \alpha_{11}l_1 + \alpha_{12}l_2 + \alpha_{13}l_3$$
$$B_2 = \alpha_{21}l_1 + \alpha_{22}l_2 + \alpha_{23}l_3$$
$$B_3 = \alpha_{31}l_1 + \alpha_{32}l_2 + \alpha_{33}l_3$$

or, more compactly,

$$B_i = \alpha_{ij} l_j \qquad i,j = 1, 2, 3$$

where α_{ij} is called the "state of dislocation tensor," and the Einstein summation convention is observed.

Since α_{ij} is invariant under a change of coordinates, it can be referred to the coordinate axes x_i. Then the α_{jj} refer to pure screw dislocations lying parallel and perpendicular to the x_i, whereas the α_{ij} refer to edge dislocations.

To relate lattice curvature to the state of dislocation tensor, we begin by defining the curvature tensor. Small rotations, as indicated in Fig. 4.40, can

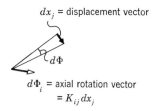

dx_j = displacement vector

$d\Phi_i$ = axial rotation vector
$= K_{ij} dx_j$

EXAMPLES

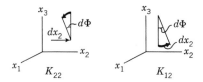

Fig. 4.40 Definition of the components of the curvature tensor above; some examples below.

be described by means of displacement vectors dx_j and axial rotation vectors $d\Phi_i$. These two vectors are related by

$$d\Phi_i = K_{ij} dx_j \qquad i,j = 1, 2, 3$$

where the coefficients K_{ij} are components of the curvature tensor. As illustrated by the examples in Fig. 4.40, components of the form K_{ii} describe twists, and components of the form K_{ij} describe rotations. Given a curvature tensor defined as above, we wish to consider an arbitrary Burgers circuit enclosing unit area and lying normal to the Ox_1 axis. We shall consider similar circuits lying normal to Ox_2 and Ox_3. Figure 4.41 shows the situation we are considering, where the Burgers circuit is at $ABCD$ and the closure

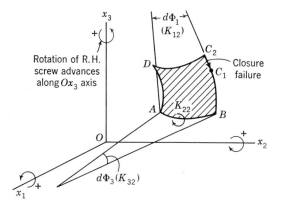

Fig. 4.41 General Burgers circuit lying perpendicular to Ox_1.

failure is the short vector C_1C_2. We take point A of the circuit to be fixed and move along AB towards C_1. Since the segment AB lies parallel to the axis Ox_2, the displacements dx_1 and dx_3 are zero. So the rotations of the point B relative to the fixed point A are

$$d\Phi_1 = K_{12}\,dx_2$$
$$d\Phi_2 = K_{22}\,dx_2$$
$$d\Phi_3 = K_{32}\,dx_2$$

These rotations cause small displacements of the point C_1. For example, consider the displacement component du_3 which can be deduced with the aid of Fig. 4.42. Since AB has unit length, it has a magnitude $\frac{1}{2}K_{12}$. Similarly, the displacement in the x_2 direction caused by the rotation of $d\Phi_1$ equals K_{12} in magnitude but swings point C_1 backward and so has a negative sign. In

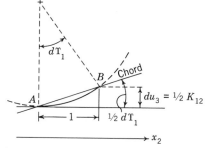

Fig. 4.42 Showing the relation between displacement in the x_3 direction and the curvature component K_{12}.

this way, the set of displacement components is

$$du_1 = K_{22} - \frac{K_{32}}{2}$$

$$du_2 = -K_{12}$$

$$du_3 = \frac{K_{12}}{2}$$

In a similar way, motion along the line BC generates displacements at C_1 given by

$$du_1' = \tfrac{1}{2}K_{23}$$
$$du_2' = -\tfrac{1}{2}K_{13}$$
$$du_3' = 0$$

Thus, the total displacements of C_1 are

$$K_{22} - \tfrac{1}{2}K_{32} + \tfrac{1}{2}K_{23}$$
$$-K_{12} - \tfrac{1}{2}K_{13}$$
$$\tfrac{1}{2}K_{12}$$

Also, it can be shown that the total displacements of C_2 are

$$\tfrac{1}{2}K_{23} - K_{33} - \tfrac{1}{2}K_{32}$$
$$-\tfrac{1}{2}K_{13}$$
$$K_{13} + \tfrac{1}{2}K_{12}$$

Then the closure-failure vector B_i, which is the difference in displacements of the points C_1 and C_2, has the following components:

$$B_1 = -(K_{22} + K_{33})$$
$$B_2 = K_{12}$$
$$B_3 = K_{13}$$

But, as was shown previously,

$$B_i = \alpha_{ij}l_j$$

and in this case $l_j = 1,0,0$ so the components of B_i are

$$B_1 = \alpha_{11} = -(K_{22} + K_{33})$$
$$B_2 = \alpha_{21} = K_{12}$$
$$B_3 = \alpha_{31} = K_{13}$$

so we have established relations between three of the dislocation tensor components and components of the curvature tensor.

Taking similar circuits normal to the axes Ox_2 and Ox_3 yields the following additional relations:

$$\alpha_{12} = K_{21} \qquad\qquad \alpha_{13} = K_{31}$$
$$\alpha_{22} = -(K_{33} + K_{11}) \qquad \alpha_{23} = K_{32}$$
$$\alpha_{32} = K_{23} \qquad\qquad \alpha_{33} = -(K_{11} + K_{22})$$

and if all of the terms are collected, the curvature tensor becomes

$$\begin{bmatrix} -(K_{22} + K_{33}) & K_{21} & K_{31} \\ K_{12} & -(K_{33} + K_{11}) & K_{32} \\ K_{13} & K_{23} & -(K_{11} + K_{22}) \end{bmatrix}$$

or, in concise notation,

$$i,j,k = 1, 2, 3$$
$$\alpha_{ij} = K_{ji} - \delta_{ij}K_{kk} \qquad \delta_{ij} = \begin{matrix} 1 & \text{if } i \neq j \\ 0 & \text{if } i \neq j \end{matrix}$$

Inversion of this equation yields

$$K_{ij} = \alpha_{ji} - \tfrac{1}{2}\delta_{ij}\alpha_{kk}$$

and we see that if the state of dislocation tensor is known, then the lattice curvature can be found everywhere. If we are dealing with discrete dislocations, this is an average curvature; but if the dislocations are continuously distributed, it is a tensor curvature field.

4.10 DISLOCATION LINES IN NONCRYSTALLINE SOLIDS

Dislocation lines take a particularly simple form in crystals, because the Burgers vector is fixed for a given line and is restricted to a few possible values that are determined by the crystal structure. In a noncrystalline solid (silica glass being the prototype), a dislocation line will have a somewhat variable Burgers vector along its length, as suggested in Fig. 4.43.

The drawings are projections of the positions of the silicon atoms in a single sheet onto the plane of the drawing. The oxygen atoms are not shown, but each silicon atom is bonded to an oxygen atom that lies just above it plus three that lie just below it parallel to the plane of the drawing. If parts of the upper oxygen layers are translated while their remainders are not, dislocation lines can be formed that are indicated by the dashed lines in the figure. The arrows represent the translations that move oxygen atoms in the next layer up from initial sites to equivalent final sites during an elementary motion of the dislocation line.

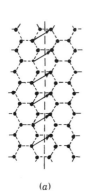

(a)   (b)

Fig. 4.43   Dislocation lines in crystalline and noncrystalline silica compared (only silicon atoms are
drawn): (a) plane of crystal; (b) projection of sheet
of glassy network.

In the noncrystalline case at the right of Fig. 4.43, the mean Burgers
displacement has a definite value that is determined by the network dimensions, but there will be fluctuations in both its magnitude and its direction
along the line. In order to minimize the energy of such a dislocation, it is
necessary for the mean **b** to be conserved over long distances, so although
the local **b**'s may fluctuate, there should be long-range correlations (if one
allows occasional large energy densities, even this condition can be relaxed).
Furthermore, there will be little tendency for the line to lie on a single plane,
and its local structure will change as it moves. Nevertheless, it is expected
that such dislocations will exist in noncrystalline solids, especially under flow
conditions. When they are viewed with a somewhat fuzzy microscope (resolution 10 Å), their behavior should resemble that of dislocations in crystals.

4.11 THE CORES OF DISLOCATIONS

Although it is often convenient to think of dislocations as "lines," they
really have more of the character of flexible rods. That is, they have a finite
extent (determined by their elastic fields) and an internal structure with a
minimum "graininess" determined by the size of the atoms that are involved.

Up to this point in the present discussion, dislocations have been characterized only by their Burgers vectors. However, in real crystals a dislocation
contains a central region whose size and configuration is dependent on the
particular structure of the crystal in addition to the Burgers vector. Furthermore, dislocations with arbitrary Burgers vectors are very unlikely to arise

in crystals, because the shear associated with them would leave noncrystal-line material behind. This would raise the energy of the system more than could be compensated by the work done during the dislocation's motion. In contrast, when the Burgers vector is such that crystal structure is restored behind the dislocation, almost all the plastic work done during a movement of the dislocation contributes to a decrease in the potential energy of the system.

PERFECT DISLOCATIONS

There are two general deformation modes which restore the crystal structure following a deformation. These are known as *translation-gliding* and *twin-gliding*. Each of them is illustrated in Fig. 3.4. During translation-gliding, one part of a crystal glides over another part by an amount equal to one unit of the structural pattern (or an integral multiple thereof). The dislocations that facilitate this process have Burgers vectors equal to the primitive translation vectors of the crystal structure. Also, since the energy of a dislocation is proportional to the square of the Burgers displacement (this will be demonstrated later), there is a strong tendency for glide dislocations to have as Burgers vectors the *smallest* translation vector of the crystal structure.

During the second type of deformation, called twin-gliding, one part of a crystal is transformed into the mirror image of the remainder of the crystal (as in Fig. 3.4). This requires dislocations with particular displacement directions and magnitudes. It also requires that there be a plane in the structure that is not initially a plane of mirror symmetry, but becomes one during the deformation process. Such dislocations are called *twinning dislocations*.

Dislocations that restore the original crystal structure after they pass a particular point are called *perfect dislocations*.

IMPERFECT DISLOCATIONS

There are many possible translation vectors in a crystal that do not restore the structure. Dislocations which have them as Burgers vectors are called *imperfect*. The best-known examples are the *partial* dislocations that can exist between close-packed layers of atoms. Many different crystal structures can be constructed by stacking close-packed layers in various patterns. Therefore, sandwiches of two or more such layers are prevalent structural elements in crystals. The possibility for partial dislocations arises because the best way to translate an atom in one close-packed layer that rests on another is not a direct translation from one site to an equivalent one. This causes a relatively large increase in specific volume during the process. Instead, it is advantageous for the atom to make a series of two partial translations which add vectorially to make the complete translation. This can be best understood by actually moving a sphere around as it rests on a substrate of close-packed spheres. Then, as illustrated in Fig. 4.44, it is

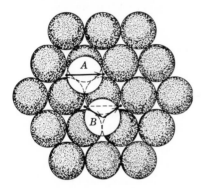

Fig. 4.44 Comparison of midglide positions for spheres resting on a close-packed substrate and translating from one position to an equivalent one.

found that if the sphere is moved directly from one site to the next, it passes through an intermediate position where it is mechanically unstable relative to the substrate, and it is lifted to a relatively high position in this unstable state. On the other hand, if it is moved in a series of two partial translations, its intermediate position lies at a natural saddle point between two spheres of the substrate, and it is not lifted so high above the substrate as in the previous case. The second process has two advantages over the first: one is that the structure is not so severely disrupted during the translation; and second, the Burgers displacements of the two partial dislocations are significantly shorter than the displacement of the unit dislocation. They are sufficiently short that the sum of their squares is less than the square of the single large one. This means that they have reduced elastic energy (this will be demonstrated later).

A single partial dislocation moving through a crystal leaves a structural fault behind. This is illustrated in Fig. 4.45, which shows the distortion of a schematic structure caused by a partial translation. The structural fault that is created by the partial dislocation can be defined by a fault vector $\mathbf{f}$; and the partial-dislocation rod is centered at the boundary of the faulted

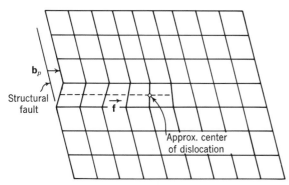

Fig. 4.45   Partial translation of one part of a crystal creates a structural fault **f**. An imperfect (partial) dislocation lies at the boundary of the faulted plane.

surface. Since the faulted region has a structure that is different from the normal crystal structure, its specific energy is greater. This increased energy per atom can range from the binding energy of the crystal down to $\frac{1}{10}$ percent (approximately 10 ergs/cm²) or less of this. The partial dislocations that are actually observed in crystals tend to have low-energy faults associated with them.

In monatomic crystals, the most common type of partial dislocation has its Burgers vector parallel to the fault vector and forms by gliding motions. This type is called the *Shockley partial dislocation*. Its atomic displacements have already been illustrated in Fig. 4.44 and in a somewhat different way in Fig. 4.46. This figure shows how a partial dislocation changes the stacking sequence behind it in a normally face-centered-cubic structure.

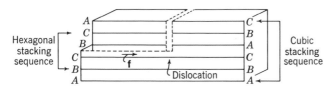

Fig. 4.46   Change of stacking sequence for close-packed atomic layers caused by passage of a Shockley partial dislocation into a crystal.

In a monatomic close-packed structure it is also possible to form partial dislocations whose Burgers vectors are not parallel to the fault vector. These are called *Frank partial dislocations*, and they can be formed only by insertion or removal of layers of atoms, and not by means of gliding alone. Two possibilities arise: one, if part of an atomic layer is removed from the structure, as illustrated at the left in Fig. 4.47; or on the other hand, if material is inserted, as at the right in Fig. 4.47. These are said to be of the negative and positive types, respectively. Comparison of the drawings of the two

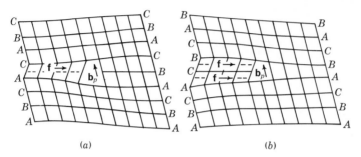

(a)   (b)

Fig. 4.47 Frank partial dislocations with faults perpendicular to Burgers vectors: (a) part of B layer removed to form negative (intrinsic) Frank partial; (b) part of C layer removed to form positive (extrinsic) Frank partial.

cases shows that distinctly different local structures arise. When part of a layer is removed, a single layer of the twin orientation forms; whereas when part of a layer is inserted, two layers of twinned orientation form. Also, since these partials cannot be formed via glide, neither can they glide after they have been formed (they can only move by means of diffusion of atoms to or from the half plane).

In polyatomic crystal structures many possible partial dislocations can be formed. Most of the possibilities will not actually be observed because they cause too much disruption of the crystal structure, but a few have sufficiently small fault energies associated with them to allow them to be metastable and hence observable. Perhaps the simplest is the one illustrated in Fig. 4.48, which shows a simple structural arrangement of two atomic species. An example of this structural arrangement is the cube face of a sodium chloride crystal. As the illustration indicates, it is possible for a unit dislocation in this structure to split into two partials which have the same direction as the unit dislocation but only half its length. Between the two partial dislocations the structure is faulted, so the arrangement of the two atomic species differs from the normal arrangement and atoms of the same kind become juxtaposed. In metal crystals, since the cohesive energy is relatively insensitive to the detailed arrangement of the atoms (it depends mainly on the packing density), a fault like the one in Fig. 4.48 does not

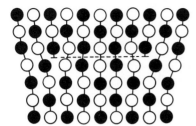

Fig. 4.48 Schematic arrangement of atoms near an edge dislocation in a diatomic crystal structure. Note the two partial dislocations separated by a section of antiphase domain boundary.

cause much increase in energy, and so it may exist in a real crystal. On the other hand, in an ionically bonded crystal, the juxtaposed ions would strongly repulse one another, causing a large increase of energy and making the existence of such a fault quite unlikely.

SOME PARTICULAR CRYSTAL STRUCTURES

The fact that dislocations in real crystals cannot be characterized by their Burgers vectors alone is demonstrated particularly clearly by the behavior of crystals with the *rock salt structure*. In this structure, dislocations with the same Burgers vector ($\langle 110 \rangle$) can move on at least two different planes, $\{001\}$ or $\{1\bar{1}0\}$, as indicated in Fig. 3.8. However, the structures at the centers of the dislocations are significantly different, as suggested in Fig. 4.49.

In the case of a $\langle 110 \rangle$ dislocation lying on an $\{001\}$ plane, when the ions in the plane just above the glide plane have been translated one-half of the unit translation distance they are brought into the configuration shown in Fig. 4.49d. Study of the figure demonstrates that no nearest-neighbor pairs of ions have opposite charge signs in this position; they all have like signs. Therefore, the binding of the crystal is locally destroyed by the translation, and this type of dislocation does not tend to form. Dislocations with this same Burgers vector have lower energy if they lie on $\{1\bar{1}0\}$ planes. Then a partial translation of the upper layer of ions leaves oppositely charged ions in nearest-neighbor positions (Fig. 4.49b), and cohesion remains across the glide planes.

The simple geometric considerations above are consistent with the behavior of ionic crystals such as MgO, LiF, NaCl, and KCl. Some crystals have this same atomic structural arrangement, but do not have their valence

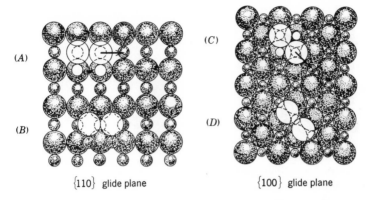

Fig. 4.49 Comparison of initial (*a*) and midglide (*b*) positions for glide in the $\langle 110 \rangle$ direction on the $\{100\}$ and the $\{110\}$ planes of the rock salt structure. Note that the $\{100\}$ plane is close-packed (and hence more smooth), but a high-energy electrostatic fault exists for its midglide position.

electrons localized. For them, since the $\{001\}$ planes are "smoother" and more widely spaced than the $\{1\bar{1}0\}$ ones, dislocations prefer to lie on the former set. Examples are crystals of PbTe and ionic crystals such as NaCl at high temperatures.

The example of the rock salt structure also serves to disprove the idea (at one time quite prevalent) that the preferred glide system in a crystal is the one for which the *glide shear* is a minimum. The glide shear is the Burgers displacement divided by the glide-plane spacing. For the preferred $\{110\}\langle 1\bar{1}0 \rangle$ system in crystals of the rock salt type, this quantity is 2, which is considerably greater than its value of $\sqrt{2}$ for the $\{001\}\langle 1\bar{1}0 \rangle$ system.

An almost completely general rule is that the preferred Burgers vector is the shortest possible translation vector of the crystal lattice. Since the elastic energy of a dislocation increases with $|\mathbf{b}|^2$, this might be expected. However, it is not strictly obeyed because the energy of the core of a dislocation is not always a minimum when $|\mathbf{b}|^2$ is a minimum. An example of maverick behavior is provided by the compound Ni_3Al which has the Cu_3Au structure (sketched in Fig. 4.50). The shortest translation vector in this structure is $a\langle 001 \rangle$, yet the observed glide direction is $a\langle 110 \rangle$. This "anomalous" behavior occurs because the $a\langle 110 \rangle$ dislocation can dissociate into two $(a/2)\langle 110 \rangle$ dislocations plus a structural fault of relatively low energy. It appears that the mobility of this dissociated dislocation is less than that of a unit one of the $a\langle 001 \rangle$ type.

In monatomic crystals, the structure at the center of a dislocation may

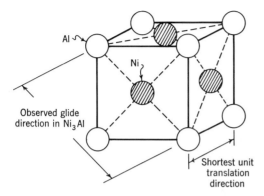

Fig. 4.50 Glide direction in the ordered Ni₃Al structure.

also depend on its glide plane, especially in crystals with "layered" structures such as beryllium, graphite, and zinc. All of these belong to the hexagonal crystal system and are bonded more tightly parallel to the hexagonal layers than perpendicular to them. They tend to glide in close-packed directions that lie in the hexagonal planes. The ease with which they do this depends sensitively on the plane across which the gliding displacement takes place. Gliding occurs easily between the weakly bonded hexagonal planes (this being the basis of the lubricating qualities of graphite), but with more difficulty along the prismatic planes which are perpendicular to the hexagonal ones.

A conclusion of this section is that in addition to a Burgers vector and a tangent vector, dislocations in real crystals require the glide plane to be given as part of their specification.

SOLID SOLUTIONS

During the movement of a dislocation through a solid solution, two things happen at its core. One is that the structure of the core changes somewhat when the dislocation passes over a solute atom. The other is that the coordination number of the solute changes when it lies at the core of a dislocation. These effects can lead (under particular circumstances) to a marked variation in the energy of the dislocation's core in the vicinity of a solute atom, tending to restrict its motion. Especially marked variations will occur in the case of solute atoms whose electronic structure favors either a particular coordination number or an atomic environment with a particular symmetry. There are a great many specific instances that can be imagined for these effects, but the discussion here will be restricted to one of the most important, as an example. This is the case of steel, for which iron is the matrix and carbon is an interstitial solute. With minor modification, the discussion also applies to other body-centered-cubic matrices and to other interstitial solutes.

The {110} glide plane of an iron crystal is sketched schematically in Fig. 4.51, together with some atoms of the adjacent plane above it. For hard spheres, the set of three displacements that accomplish a unit ⟨111⟩ translation with a minimum of disturbance at the glide plane is sketched.

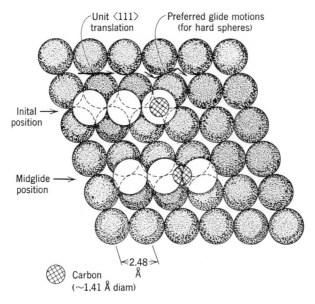

Fig. 4.51   Glide motion on the {100} planes of iron and its effect on the local environments of dissolved carbon atoms.

The effect of dissolved carbon may be seen by comparing the sketch of three atoms above the glide plane in their initial positions with the sketch that depicts their midglide positions. In the former sketch an interstitial carbon atom is indicated which lies in a pseudo-octahedral hole giving it an average coordination number of 6 (4 plus 2). However, when the iron atoms above the glide plane are translated to their midglide positions, the hole is no longer even approximately octahedral, and the coordination number becomes approximately 7.

It is clear from the structure of iron carbide (Fe_3C) and various other transition metal carbides that in such matrices carbon prefers a coordination number of 6 (sometimes in an octahedral hole, and sometimes in a prismatic one). Therefore, the energy of a carbon atom is increased significantly in the midglide position within a dislocation core. This creates substantial resistance to its motion.

Dissolved carbon also prevents the iron atoms in its vicinity from gliding in the preferred directions by blocking the most compatible positions. Thus carbon also increases the energies of iron atoms near the center of a dislocation.

The geometric properties of iron-carbon complexes as discussed above indicate qualitatively how carbon markedly influences the plasticity of iron. The configurations are sufficiently nonsymmetric, however, to make it difficult to develop a quantitative atomic theory of the effects. Most of the models that have been proposed cannot survive sharp criticism and have little predictive power.

If a solid solution is not completely random but possesses some short-range order (an excess number of solute pairs or antipairs), then a dislocation passing through it will tend to decrease the amount of short-range order (Fisher, 1954). Since this will increase the crystal's energy, it creates resistance to dislocation motion.

4.12 ATOMIC STRUCTURE OF KINKS AND JOGS

In a homogeneous isotropic medium, kinks and jogs on dislocation lines can be quite wide, so the transition of the dislocation from one path to another occurs gradually. However, real crystals are never homogeneous and seldom isotropic, so kinks and jogs whose offsets are of atomic dimensions tend to be localized and so become a special class of "point defects." They play an essential role in the motion of dislocations, because the elementary processes occur at them. Thus, the absorption or emission of vacancies (or interstitials) at jogs results in the "climb" of an edge dislocation, and the motion of kinks results in translation-gliding.

Kinks and jogs on dislocations can arise in a variety of ways. Whenever a dislocation line moves, for example, kinks must be made along it quite profusely, since segments of it advance in multiples of the Burgers vector. Cross glide produces either kinks or jogs, depending on the plane on which it occurs, and the intersection of moving dislocations is another means for jog and kink production. When one dislocation intersects another, the first dislocation becomes jogged by an amount equal to the component of the Burgers vector of the intersecting dislocation that lies perpendicular to the glide plane of the first dislocation. This latter will also become kinked by an amount equal to the Burgers vector component that lies parallel to its glide plane.

Jogs have a special importance in ionic crystals because they are sometimes charged. This was first pointed out by Seitz (1951), and it has been discussed in further detail by Frank (1958). In the rock salt structure, jogs may be either neutral or charged on $\{110\}\langle110\rangle$ dislocations. This may be seen by considering an extra $\{100\}$ half plane associated with a $\langle110\rangle$ edge dislocation (Fig. 4.52). The edge of the half plane may jog by either an odd number of

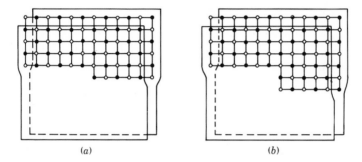

(a)   (b)

Fig. 4.52 Jogs on $\{100\}$ half planes associated with $\{110\}$ $\langle1\bar{1}0\rangle$ dislocations in the rock salt structure: (a) charged jog; (b) neutral jog.

ion rows (Fig. 4.52a) or an even number (Fig. 4.52b). In the first case there is one uncompensated ion left at the jog, so it acquires a charge of $\pm q/2$ where q is the ionic charge. In the second case, there is an unbalanced dipole at the end of the jog, but no extra charge.

Jogs are more simply defined, in general, than kinks, because an abrupt change does not necessarily occur at a kink. In the case of an extended dislocation it is especially difficult to produce a sudden curvature of the ribbon-like dislocation. One case that does allow sharp kinks is that of covalent crystals which have the diamond, zinc blende, or wurtzite structures. For this case the atoms have distinctly lower energies when they lie at the centers of their tetrahedral coordination shells. Therefore, the gradual transition of a dislocation core from one path to another is not favored, because it distorts many of the tetrahedral angles. A localized sharp transition distorts only a few angles, causing a smaller energy increase. This would not be true if the distortion energy changed linearly with distortion angle, but it does not.

It is difficult to draw a clear illustration of the atomic structure near a kink, but Fig. 4.53 is an attempt to do this. It indicates how a line of "dangling bonds" shifts from one row to an adjacent parallel row.

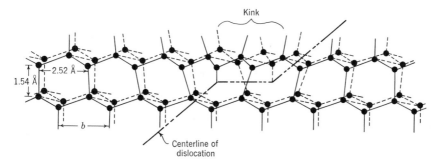

Fig. 4.53   Atomic configuration near a kink along a dislocation line in the diamond structure. The glide plane is {111}, and the Burgers vector is ⟨01$\bar{1}$⟩. The line tangent lies 60° away from the Burgers vector.

4.13 TWINNING DISLOCATIONS

It may be seen that the faulted region associated with the partial dislocation at the right in Fig. 4.47 has mirror symmetry with respect to the remainder of the crystal. Thus partial dislocations are sometimes also twinning dislocations, although this is not necessarily the case. A more explicit illustration of a twinning dislocation is presented in Fig. 4.54. Here the boundary be-

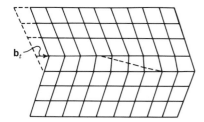

Fig. 4.54   Schematic drawing of a dislocation at the boundary between the two regions of a twinned crystal. The associated Burgers vector is $\mathbf{b}_t$.

tween the twinned and untwinned portion of the crystal constitutes a dislocation line with the Burgers vector $\mathbf{b}_t$.

Twinning dislocation lines have many of the same geometrical properties as have already been discussed for translation dislocations. However, their description is somewhat more complicated because they lie at the junction between two differently oriented structures. Therefore, a formal description must be referred to a reference lattice which can be chosen as either the twin or the matrix or as an intermediate lattice that is different from either.

The production of a macroscopic amount of twinned material requires some means for the regeneration of twinning dislocations because freshly grown or annealed crystals do not normally contain large concentrations of them. The various regeneration mechanisms that have already been discussed are possible for twinning dislocations, providing they can result in a correlated motion of the dislocation from one plane to the next. This is necessary because, unlike the case of translation-gliding, each succeeding plane must be shifted in some systematic fashion in order to generate a macroscopic twin.

A regeneration mechanism that has not been mentioned as yet, and which is capable of causing correlated twinning on successive planes, is illustrated in Fig. 4.55. Here a twinning dislocation with Burgers vector $\mathbf{b}_t$ is connected

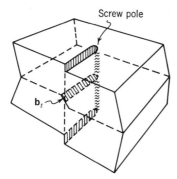

Fig. 4.55 Pole mechanism for continuous twinning through the motion of a twinning dislocation along the helical ramp created by a screw-dislocation "pole."

to a screw dislocation threading the twinning plane. The screw dislocation's Burgers vector has a component equal to the twin-plane spacing. Therefore, each time the twinning dislocation turns around the screw dislocation it moves up to a successive plane where it can make another turn. This is known as the Cottrell-Bilby pole mechanism.

4.14 CLEAVAGE DISLOCATIONS

The half plane of atoms that is associated with an edge dislocation causes a dilation of the region that lies just below the extra half plane. This leads to a

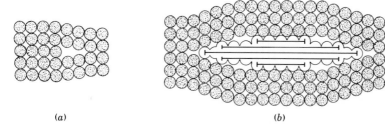

Fig. 4.56 Dislocations as elementary cracks: (a) the microscopic crack at an edge dislocation; (b) a distribution of virtual edge dislocations that is equivalent to a crack.

configuration that is very similar to a microscopic crack (Fig. 4.56a). Therefore, edge dislocations can be used in a formal way to represent cracks. This may be done, as shown at the right in Fig. 4.56b, by adding together an appropriate distribution of virtual edge dislocations. The properties of such a distribution, especially the elastic ones, are very similar to the properties of a crack, and this representation is sometimes convenient to use, because it allows a crack to be broken down into more elementary parts. This is especially true, for example, in thinking about and analyzing the behavior that can be expected at the very tip of a crack, and the technique has been used to good advantage beginning with a paper by Friedel (1959).

4.15 DISLOCATION REACTIONS

There are a number of possible events in which the Burgers vectors of dislocations change through some kind of interaction. Events of this type are outlined below.

ANNIHILATION

Two parallel dislocation lines having Burgers vectors of the same magnitude and direction but opposite sign may interact in such a way as to cancel one another, or so as to nearly completely cancel leaving behind a row of defects. These two cases will be described in turn.

1. *Complete Annihilation.* In this case, the vector sum of the reacting Burgers vectors is zero:

$$\mathbf{b} + (-\mathbf{b}) \rightarrow 0$$

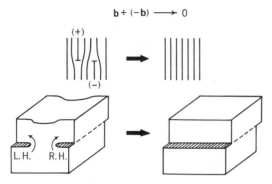

Fig. 4.57 Annihilation of dislocations having Burgers vectors with opposite signs.

The parallel dislocations may be of either the edge or screw type, as illustrated in Fig. 4.57.

2. *Partial Annihilation.* If two parallel dislocation lines of the edge type with opposite Burgers vectors and lying on adjacent glide planes react, they are drawn together by strong attractive forces but do not disappear when they meet. Instead, depending on the disposition of their extra half planes, a row of vacancies or interstitials remains after the reaction. Thus the reaction may be written

$$\mathbf{b}_1^a - \mathbf{b}_2^{a \pm b} \rightarrow \begin{pmatrix} \text{vacancy row} \\ \text{interstitial row} \end{pmatrix}$$

and the geometry is sketched in Fig. 4.58. In monatomic crystals, simple lines of vacancies or interstitials are the consequences of this kind of reaction, but in diatomic crystals the reaction product may be either molecular vacancy interstitial rows or atomic vacancy interstitial rows.

ELASTIC INTERACTIONS

If two parallel edge dislocations that lie on nearby (but not immediately adjacent) glide planes move together, they can interact through their elastic stress fields to form various elastically bound complexes. If they lie on parallel glide planes, the result can be a simple (homogeneous) dipole. They can also interact elastically, however, if their glide planes are not parallel as indicated in Fig. 4.59, to form a heterogeneous dipole as contrasted with the homogeneous case above.

These elastically bound complexes cannot arise for screw dislocations in isotropic crystals, because the dislocations will simply move together along

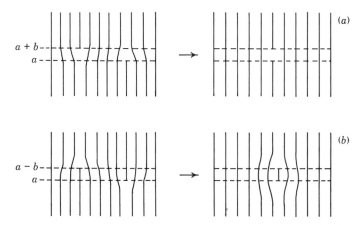

Fig. 4.58 Schematic reactions between dislocations that have Burgers vectors of opposite signs and which lie on adjacent planes: (a) a vacancy row is produced; (b) an interstitial row is produced.

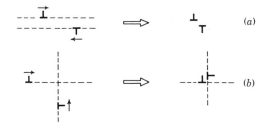

Fig. 4.59 Formation of elastically bound dislocation pairs: (a) homogeneous dipole; (b) heterogeneous dipole.

arbitrary planes until they mutually annihilate. In anisotropic crystals, on the other hand, screw dislocations have preferred glide planes, and if they tend to dissociate appreciably on these planes, they will strongly prefer to stay on them. Pairs of screw dislocations with restricted glide planes can interact elastically just as the edge dislocations described above do to form dipoles.

STRUCTURAL INTERACTIONS

Parallel dislocation lines with different Burgers vectors can combine with each other in a variety of ways. Such reactions may be conveniently described by means of vector equations, and some of the possibilities are listed below:

Combination: $\mathbf{b}_1 + \mathbf{b}_2 \rightarrow \mathbf{b}_3$
Decomposition: $\mathbf{b}_1 + \mathbf{b}_3 \rightarrow \mathbf{b}_3 + \mathbf{b}_4$
Dissociation: $\mathbf{b}_1 \rightarrow \mathbf{b}_p + \mathbf{b}_p' + \mathbf{f}$

The first of the above three equations describes the reaction between two dislocations to form a third one; the second describes the interaction of two dislocations to form two other dislocations; and the third is of the dissociation type, where one dislocation splits into two or more partial dislocations plus a stacking fault.

There are very many reaction possibilities—far too many for a thorough description here. Only two cases will be outlined as typical examples: one of these occurs with considerable regularity in face-centered-cubic metals, and the other occurs in crystals with the rock salt structure.

LOMER-COTTRELL REACTION IN FACE-CENTERED-CUBIC METALS. Figure 4.60 illustrates what happens when two dislocations with Burgers vectors of the $\langle 10\bar{1} \rangle$ type meet as they move along intersecting glide planes of the

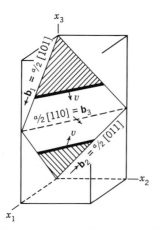

Fig. 4.60   Lomer-Cottrell reaction between two $\langle 10\bar{1} \rangle$ dislocations in a cubic crystal.

$\{111\}$ type in the face-centered-cubic structure. The two dislocations react to form a third with a Burgers vector that belongs to the $\langle 10\bar{1} \rangle$ class but whose glide plane is of the $\{001\}$ type instead of the normal $\{111\}$ type.

Since this glide plane is not a primary one in this structure, the resulting dislocation is not expected to glide readily. The corresponding reaction equations are listed below:

$$\mathbf{b}_1^a + \mathbf{b}_1^b \rightarrow \mathbf{b}_1^c$$
$$\frac{a}{2}[10\bar{1}] + \frac{a}{2}[011] \rightarrow \frac{a}{2}[110]$$

KEAR-PRATT REACTION IN CRYSTALS OF THE ROCK SALT TYPE.   In the rock salt structure, two primary glide planes of the {110} type which intersect one another can often be activated in a simple tension or compression test. Dislocations that glide simultaneously on these planes will have different Burgers vectors, as indicated in Fig. 4.61, both belonging to the class $\langle 01\bar{1} \rangle$.

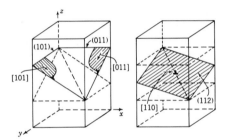

Fig. 4.61   Illustration of the Kear-Pratt dislocation reactions in the rock salt structure.

When the two dislocations meet at the intersection between the two planes, the following reaction will occur:

$$\mathbf{b}_1^a + \mathbf{b}_1^b \rightarrow \mathbf{b}_1^c$$
$$\frac{a}{2}[0\bar{1}1]_{101} + \frac{a}{2}[10\bar{1}]_{011} \rightarrow [1\bar{1}0]_{112}$$

The Burgers vector of the resulting dislocation belongs to the same class as those of the reactants, but it is of the edge type with a {112} glide plane. Since this is not a primary glide plane, the reaction product cannot move readily. A result of this lack of mobility is that the reaction causes substantial hardening of ionic crystals if it occurs with regularity during plastic deformation.

DISSOCIATION REACTIONS

The reaction equation for dissociation may be written

$$\mathbf{b} \to \mathbf{b}_p + \mathbf{b}_p + \mathbf{f}$$

and this can be generalized quite simply as follows:

$$\mathbf{b} \to \sum_i^n \mathbf{b}_p^i + \sum_j^m \mathbf{f}_j$$

Some representative dissociation reactions are outlined below.

HEXAGONAL-CLOSE-PACKED STRUCTURE. Figure 4.62a shows schematically how a unit dislocation dissociates in this structure. It simply separates along the basal plane (which is often the primary glide plane). This decreases

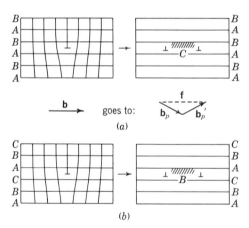

Fig. 4.62 Dissociation reactions in close-packed monatomic crystals. (a) Hexagonal-close-packed structure; (b) face-centered-cubic close-packed structure.

the intensity of the strains in the immediate vicinity of the dislocation and also creates a stacking fault as indicated by the hatching in the figure. If the dislocation lies between two close-packed planes of the type AB, then a layer of the C type is produced at the center of the dislocation as is sketched in the figure.

FACE-CENTERED-CUBIC STRUCTURE. This case is quite similar to that of the hex. c.p. structure because both structures are built up by stacking close-packed layers on top of one another (Fig. 4.62b). The stacking sequence for the f.c.c. structure is ABC-ABC, compared with the simple AB-AB sequence in the hex. c.p. case. If for example then, an edge dislocation has its glide plane between two layers of the A and C type, it will create a short section of the B type when it dissociates, as is sketched in Fig. 4.62b.

DISSOCIATION OF LOMER-COTTRELL DISLOCATIONS IN THE FACE-CENTERED-CUBIC STRUCTURE. The dislocation that results from the Lomer-Cottrell reaction described in the previous section may dissociate as indicated schematically in Fig. 4.63. It can become three partial dislocations plus two

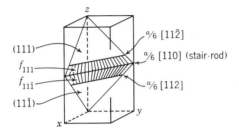

Fig. 4.63 Possible dissociation of Lomer-Cottrell dislocation.

stacking faults. The reaction equation that describes this splitting is

$$\left(\frac{a}{2}[110]_{100}\right)_1 \rightarrow f_{111} + f_{11\bar{1}} + \left(\frac{a}{6}[11\bar{2}]_{111}\right)_2 + \left(\frac{a}{6}[112]_{11\bar{1}}\right)_3 + \left(\frac{a}{6}[110]_{100}\right)_4$$

It may be seen that dissociation results in the production of two dislocations of the $\langle 11\bar{2}\rangle$ class and these two dislocations move away along different $\{111\}$ planes. They leave behind a small dislocation of the type $\langle 110\rangle$. The relative magnitudes of the Burgers vectors involved are

$$(|\mathbf{b}|_{110})_1 = \frac{1}{\sqrt{2}}$$

$$(|\mathbf{b}|_{11\bar{2},112})_{2,3} = \frac{1}{\sqrt{6}}$$

$$(|\mathbf{b}|_{110})_4 = \frac{\sqrt{2}}{6}$$

So the sum of the squares of the Burgers vectors on the left-hand side = 0.50, whereas it is only 0.38 on the right-hand side. Therefore, there is a net gain in elastic energy during the reaction (to a first approximation).

BODY-CENTERED-CUBIC STRUCTURE. There are a number of possible dissociation reactions in the body-centered-cubic structure, and only one will be described here as an example. The displacements of this dissociation have already been illustrated in Fig. 4.51; no additional sketch will be given here. The dislocation's Burgers vector is of the type $(a/2) \langle 111 \rangle$, and its glide plane is of the type $\{110\}$. It can dissociate into three shorter dislocations, as sketched in Fig. 4.51 and described by

$$\frac{a}{2}[\bar{1}\bar{1}1] \to \frac{a}{8}[\bar{1}\bar{1}0] + \frac{a}{4}[\bar{1}\bar{1}2] + \frac{a}{8}[\bar{1}\bar{1}0] + 3 \text{ faults}$$

Such a dissociation seems natural if one studies how a hard-sphere model of the b.c.c. structure fits together. However, it should be kept in mind that a hard-sphere model is not really appropriate for a non-close-packed structure (such as the b.c.c. one), and this dissociation may not be favored by the electronic structures of such crystals.

LAYERED SILICATES (TALC). In complex layered structures, dislocations whose Burgers vectors lie parallel to the plane of the layering can become widely dissociated. A good example is talc (a layered silicate similar in structure to mica and pyrophyllite). A very schematic outline of its structure is presented in Fig. 4.64, where the edges of the atomic layers are sketched

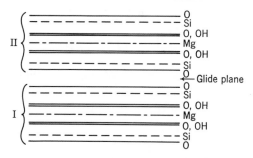

Fig. 4.64 Schematic structure of talc consisting of two magnesium silicate layers I and II that are tightly bound internally, but are only weakly bound together.

to show the stacking sequence. The overall structure is divided into two relatively weakly bound layers, and each layer has considerably more tightly bound substructure within it. Each of the two sublayers consists of silicon and oxygen on the outside which form two-dimensional sheets with internal covalent bonding. Between two of these sheets a layer of magnesium is present which is combined with hydroxyl ions. These provide hydrogen bonding between the magnesium oxide layer and the silicate layers on either side. The only bonding between the two sublayers that form the overall structure is of the Van der Waals type. Therefore, it is between the sublayers that glide dislocations form readily and can dissociate readily. Any structural faults that form in this region have quite small energies because of the weak binding.

In Fig. 4.65 one of the atomic oxygen layers that lies adjacent to the glide plane of talc is shown schematically. This layer is not close-packed, so its

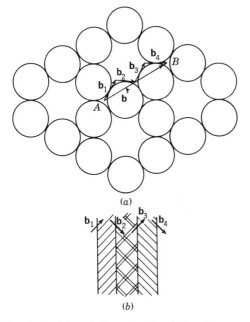

(a)

(b)

Fig. 4.65 Dissociation of glide dislocations in talc crystals: (a) arrangement of oxygen atoms adjacent to glide plane; (b) three faults created by dissociation of the Burgers vector AB.

rotational symmetry is reduced from sixfold to threefold, and this increases the size of the minimum translation vector of the structure. In a given close-packed direction, one must move from a point like A to a point like B to reach an equivalent site (instead of half this distance, as would be the case if the layer were close-packed). If the glide dislocation kept such a large Burgers vector intact, it would have a very large elastic energy, but it is able to dissociate into four partial dislocations (Fig. 4.65a). These four partial dislocations are separated by three structural faults, as sketched in Fig. 4.65b.

Extended dislocations of the above kind have actually been observed in talc crystals (Fig. 4.66) and a variety of other layered silicates by Amelinckx

Fig. 4.66 Extended dislocations in a talc crystal magnified 24,000×. (*After Amelinckx and Delavignette*. With permission from Dr. S. Amelinckx, Laboratories du C.E.N., Mol-Donk, Belgium.)

and Delavignette (1961). They are also expected to occur in relatively low-symmetry oxides such as Al_2O_3 where the oxygen ions are arranged in close-packed layers, but the aluminum arrangement reduces the overall symmetry and requires long unit translation dislocations. From the points of view of geometry and of reducing the elastic energy of a glide dislocation, one can expect fourfold dissociation to occur in Al_2O_3, as was first suggested by Kronberg (1961). On the other hand, the chemical binding between the layers of aluminum oxide is quite strong, so the structural faults created by dissociation will have much higher energies then is the case for talc, and dissociation may not actually be observed. Furthermore, at the temperatures where plastic flow has been observed in Al_2O_3, the aluminum ions are probably quite disordered. If this is the case, then the symmetry of the oxygen layers becomes hexagonal and the unit translation vector becomes shortened, reducing the driving force for dissociation.

4.16 NODAL REACTIONS

At a place where three or more dislocations meet, called a *node*, it is sometimes possible for the local energy density to be reduced by some kind of reaction. Once again, there are many possibilities and only two examples will be given.

DECOMPOSITION OF A FOURFOLD NODE

If four dislocation lines meet at a point, as sketched in Fig. 4.67a it may be possible for the lines to reduce their total length through splitting of the

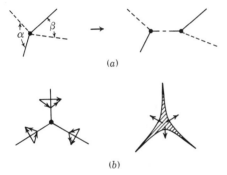

(a)

(b)

Fig. 4.67 Reactions at dislocation nodes: (a) decomposition of fourfold node; (b) extension of threefold node.

node into 2 threefold nodes connected by a short segment. This could be expected to occur, for example, if the angles α, β between pairs of the lines are less than about 70° and therefore are able to increase by means of the splitting reaction.

EXTENSION OF A THREEFOLD NODE

The three dislocations that are sketched at the left in Fig. 4.67b might be able to reduce their energies by dissociating. If they could, there would be a strong tendency for the node to dissociate as well, as sketched at the right. The size of such a node would continue to increase until the energy of the fault at the node became equilibrated with the forces caused by the curvatures of the partial dislocations at the edges of the node. Such reactions are observed in crystals that have low stacking-fault energies such as Co-Ni alloy crystals.

REFERENCES

GENERAL

Amelinckx, S.: "The Direct Observation of Dislocations," *Solid State Phys.*, suppl. 6, Academic Press Inc., New York, 1964.

Burgers, J. M., and W. G. Burgers: "Rheology," vol. 1, Academic Press Inc., New York, 1956.

Cottrell, A. H.: "Dislocations and Plastic Flow in Crystals," Oxford University Press, Fair Lawn, N.J., 1953; also, "Theory of Crystal Dislocations," Gordon and Breach, New York, 1964.

De Wit, R.: *Solid State Phys.*, **10**: 249 (1960).

Friedel, J.: "Les Dislocations," Gauthier-Villars, Paris, 1956; also, "Dislocations," Pergamon Press, New York, 1964.

Hirth, J. P., and J. Loethe: "Theory of Dislocations," McGraw-Hill Book Company, New York, 1968.

Nabarro, F. R. N.: "The Theory of Crystal Dislocations," Oxford University Press, Fair Lawn, N.J., 1967.

Read, W. T.: "Dislocations in Crystals," McGraw-Hill Book Company, New York, 1953.

Seeger, A.: *Handbuch Phys.*, **7**: 383 (1956).

Seitz, F., J. S. Koehler, and E. Orowan: in "Dislocations in Metals," AIME, New York, 1954.

—— and T. A. Read: *J. Appl. Phys.*, **12**: 100, 170, 470, 538 (1941); bibliography (Russian), "Dislocations in Crystals," Acad. Sci. U.S.S.R., Moscow, 1960.

Weertman, J., and J. R. Weertman: "Elementary Dislocation Theory," The Macmillan Company, New York, 1964.

SPECIFIC TOPICS

Amelinckx, S., and W. Dekeyser: The Structure and Properties of Grain Boundaries, *Solid State Phys.*, **8**: 325 (1959).

—— and P. Delavignette: *J. Appl. Phys.*, **32**: 341 (1961).

Benioff, H., Earthquake Source Mechanisms, *Science*, **143**: 1399 (1964).

Bilby, B. A.: Continuous Distributions of Dislocations, *Progress in Solid Mech.*, **1**: 331 (1960).

Fisher, J. C.: *Acta Met.*, **2**: 9 (1954).

Frank, F. C.: Dislocation Theory, *Nuovo Cimento*, suppl. **7**: 386 (1958).

————: The Origin of Dislocations in "Plastic Deformation of Crystalline Solids," p. 89, Department of U.S. Navy, NAVEXOS-P-834, 1950.

———— and W. T. Read: Multiplication Processes for Slow-moving Dislocations, *ibid.*, p. 44.

Friedel, J.: in "Fracture," B. L. Averbach et al. (eds.), p. 498, John Wiley & Sons, Inc., New York, 1959.

Gilman, J. J.: Dislocation Sources in Crystals, *J. Appl. Phys.*, **30**: 1584 (1959).

Kronberg, M. L.: *Acta Met.*, **9**: 970 (1961).

Koehler, J. S.: *Phys. Rev.*, **86**: 52 (1952).

Nye, J. F.: Some Geometrical Relations in Dislocated Crystals, *Acta Met.*, **1**: 153 (1953).

Orowan, E.: in "Dislocations in Metals," M. Cohen (ed.), p. 103, AIME, New York, 1954.

Seitz, F.: Speculations on the Properties of the Silver Halide Crystals, *Rev. Mod. Phys.*, **23**: 328 (1951).

Volterra, V.: *Ann. Ecole Normal Super.*, **24**: 400 (1907).

5

KINEMATICS AND DYNAMICS OF DISLOCATION MOTION

5.1 KINEMATICS

Plastic deformation is a flow process that is highly heterogeneous. The elementary flow events are associated with the motions of dislocation lines that are not uniformly distributed in a crystal. Furthermore, since dislocation lines are flexible, they can move in a variety of ways. If they were to move as rigid rods, relatively large forces would be required to act upon a given length. Therefore, it is usually more favorable for them to move by means of kinks that propagate along their lengths in order to move them transversely. Figure 5.1 shows this schematically.

KINKS AND CUSPS

The shape of a kink is determined by the static potential energy of a dislocation as it varies with position in a crystal and by dynamical factors if the dislocation is moving. An analog of the static case is a chain lying along the troughs of a sheet of corrugated iron with one end in one trough and the other in an adjacent one. The transition zone where the chain passes from one trough to the other is the analog of a kink. The intersection of the weight of the chain and the corrugated substrate causes its energy to depend on its position, as sketched schematically in Fig. 5.2 for the analogous dislocation. If the coordinates are chosen, as in Fig. 5.2, and the potential energy per unit length has the sinusoidal form

$$U = U_m - U_0 \cos \frac{2\pi y}{a} \tag{5.1}$$

then the shape of the dislocation line is determined by a balance between

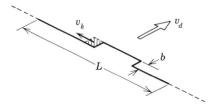

Fig. 5.1 Production of a transverse velocity v_d by longitudinal kink motions v_k along a dislocation line.

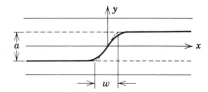

Fig. 5.2 Schematic dislocation line straddling a potential energy hill.

the tendency to minimize the amount of length with energy U_m without unduly increasing the curvatures at the two ends. It is given by Seeger and Schiller (1962):

$$y = \frac{2a}{\pi} \tan^{-1} [\exp \Gamma x] \tag{5.2}$$
$$\Gamma = \frac{2\pi}{a} \left(\frac{U_0}{U_m}\right)^{\frac{1}{2}}$$

which is approximate unless the curvature is small everywhere.

The width is defined as shown in the drawing and has the value

$$w = \frac{a}{2} \left(\frac{U_m}{U_0}\right)^{\frac{1}{2}} \tag{5.3}$$

Therefore, the kink tends to become sharp as the amplitude of the energy variation U_0 becomes large.

Even if the energy U_0 is small, kinks may exist in a dynamical sense as transverse waves analogous to those on a rope that is whipped (Laub and Eshelby, 1966).

Next, consider what happens when kinks move. If they have a constant velocity v_k, then the time t needed for one kink to traverse a length L of dislocation line is L/v_k; or the time needed for n kinks is L/nv_k. In this amount of time the dislocation moves forward a distance $= b$. Therefore, the dislocation's velocity v_d is given by

$$v_d = \frac{nb}{L} v_k$$

but n/L is just the kink density N_k (kinks per unit length), so

$$v_d = bN_k v_k \tag{5.4}$$

or

$$v_d = C_k v_k \tag{5.5}$$

where C_k is the ratio of kinks to atomic lengths (note that this equals the fractional concentration if the kinks are atomically sharp). Since $C_k \leq 1$, the kink velocity always exceeds the dislocation velocity.

This general relation can be made more specific by considering the balance that exists between the creation of kinks and their annihilation on a straight dislocation line.† Suppose that α is the number of kink pairs (one positive and one negative) that is created per second per unit length, and let λ be the mean distance between the kinks which equals $1/N_k$. Then the mean time between collisions (remembering that two kinks approach each collision site) is $\lambda/2v_k = 1/2N_k v_k$, and the collision rate is $R = 1/t = 2N_k v_k$. Then, since the pair concentration is $N_k/2$, the pair annihilation rate is

$$\frac{N_k}{2} R = v_k N_k^2$$

Using the above creation and annihilation rates, the rate of change of the kink concentration is

$$\frac{dN_k}{dt} = 2\alpha - 2v_k N_k^2 \tag{5.6}$$

When the kink density saturates, $dN_k/dt = 0$, so the expression for the saturation kink density becomes

$$N_k^* = \left(\frac{\alpha}{v_k}\right)^{\frac{1}{2}} \tag{5.7}$$

and if this is substituted into the velocity equation (5.5), the result is

$$v_d = b(\alpha v_k)^{\frac{1}{2}} \tag{5.8}$$

In order for sharp kinks to exist on a dislocation, its energy must depend markedly on the exact position of its center, otherwise it will curve smoothly

† For curved lines, kinks may flow from regions of high to low curvature.

from one place to another, as discussed above. The former situation prevails in covalently bonded crystals of the diamond or zinc blende type, whereas the latter is typical of pure metallic and ionic crystals. If the energy of a dislocation does not depend on its position relative to the matrix structure of a crystal, it nevertheless is often strongly affected by localized defects, especially by impurity atoms. These *pinning interactions* hold back short segments of moving lines, thereby causing *cusps* to form as illustrated by Fig. 5.3.

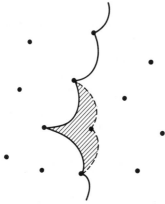

Fig. 5.3 Schematic drawing of the motion of a dislocation on a glide plane that contains many pinning points. The solid line gives the initial position.

The dislocation lines that extend back to a pinning point exert an ever-increasing force on it, until the pinned segment breaks away from it and suddenly moves ahead as the sketch of Fig. 5.3 suggests. Thus the motion is discontinuous in a way that depends on the distribution of pinning points.

The submicroscopic processes that are outlined above require detailed consideration for an atomic theory of dislocation motion. However, they cannot be studied directly by present experimental means, so speculative ideas about them do not contribute substantially to the theory of macroscopic plastic flow except by setting limits on the form that it can reasonably take. On the other hand, average dislocation velocities can be measured experimentally and therefore can be used in a quantitative flow theory, even though it is not known precisely whether a given average velocity results from small movements at a large number of submicroscopic positions or from large movements at a modest number of positions.

THE PLASTIC STRAIN RATE (ONE–DIMENSIONAL)

Because it is of basic importance, the relations between plastic flow and dislocation motions will be introduced in more than one way. First, using straight dislocation lines (one-dimensional); then using loops; and, finally, using the generalized case.

Figure 5.4a is a schematic drawing of straight edge dislocations moving through a crystalline block, and Fig. 5.4b is the same for screw dislocations.

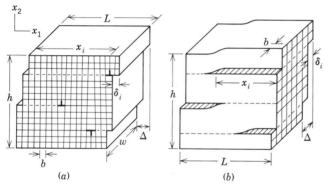

Fig. 5.4   External shape changes caused by dislocation moving through a solid: (a) a set of parallel edge dislocations; (b) a set of parallel screw dislocations.

Plastic flow changes the shape of a solid permanently, and the shape change can be quantitatively described as the displacement of one part of the solid relative to another. In this case, the displacement of the top of the block relative to the bottom is designated Δ, and to make the description of the shape change independent of the size of the body, this displacement is expressed per unit height by dividing it by h. (Note that in order to make this description compatible with the tensor description, a factor of $1/2$ must be introduced.) Then the plastic shear strain may be written

$$\epsilon_s = \frac{\Delta}{h} \tag{5.9}$$

and (using a dot to designate time differentiation) the strain rate is

$$\dot{\epsilon}_s = \frac{\dot{\Delta}}{h} \tag{5.10}$$

The overall displacement is the sum of the small displacements δ_i caused by individual dislocations which carry a local displacement b with them. When the ith dislocation starts moving across the block, the displacement associated with it necessarily equals zero. When it has moved a distance $= L$ and therefore has reached the far side of the block, the associated displacement is b. If it moves at constant velocity v, the time that this requires is L/v, so the rate of displacement is

$$\dot{\delta}_i = b\frac{v_i}{L} \tag{5.11}$$

and, if n is the total number of dislocation lines,

$$\dot{\Delta} = \sum_1^h \dot{\delta}_i = \frac{b}{L}\sum_1^h v_i \tag{5.12}$$

which can be written in terms of the average velocity $\bar{v}$:

$$\dot{\Delta} = \frac{bn\bar{v}}{L}$$

To obtain the shear strain rate, this is put into Eq. (5.10),

$$\dot{\epsilon}_s = b\bar{v}\frac{n}{hL}$$

but n/hL can be set equal to N the flux of dislocation lines through a unit cross-sectional area, so the rate equation becomes

$$\dot{\epsilon}_s = bN\bar{v} \tag{5.13}$$

Quite analogous arguments lead to the same conclusion for the screw dislocations of Fig. 5.4b.

Equation (5.13) relates the macroscopic plastic flow rate to the product of three microscopic quantities; the displacement carried by each dislocation, the concentration of dislocations, and the mean velocity of each dislocation. Thus it is analogous to other transport equations. For example, the electric current J that passes through a solid is given by

$$J = qn_cv_c$$

that is, the product of the charge q carried by each current carrier (electron, hole, or ion), the number of carriers per unit volume n_c, and the mean drift velocity of the carriers v_c.

In both cases above, the "action" per entity is large. Their effective masses are low, so their mobilities are high. Thus the flow or current is rapid even for small applied fields (stress or electric).

Equation (5.13) must be interpreted with care, because its variables are not necessarily continuous. Although N is written as a continuous variable and may be treated accordingly under some circumstances, for straight lines it is discontinuous; that is, the ith dislocation either exists or it doesn't in Fig. 5.4: it cannot partially exist. If it could, Eq. (5.13) would have an additional term of the form $b\bar{x}\dot{N}$. Similarly, the velocity $\bar{v}$ is not fundamentally continuous. The moving kinks that cause it may be of atomic dimensions, and if so, they obey the laws of quantum mechanics, not classical mechanics. Because of these factors, classical elasticity theory is useful for dealing with certain dislocation problems, but other properties require an atomic basis for their proper description.

THE PLASTIC STRAIN RATE (THREE–DIMENSIONAL)

Consider a planar dislocation loop of arbitrary shape that lies on a glide plane. Suppose that it has an initial area a and that it expands by an amount da. This will cause an external displacement $d\Delta$ in a direction parallel to the Burgers vector of the loop. By the arguments of the previous section, the magnitude of the displacement is

$$d\Delta = b\,\frac{da}{A} \tag{5.14}$$

where A is the total area of the glide plane. Then, if h is the height of the body, the differential shear strain is

$$d\epsilon = \frac{d\Delta}{h} = \frac{b}{V}\,da$$

where V is the volume of the solid. Time differentiation yields the strain rate

$$\dot{\epsilon} = \frac{b}{2V}\,\dot{a} \tag{5.15}$$

Next, suppose that a *mean loop* is being considered, that is, one with a mean size and shape. Focus on a small length element dl which has a velocity vector $\mathbf{v}$ and a unit normal vector $\mathbf{n}$ that points outward. In terms of these quantities, the rate at which the area of this loop changes is

$$\dot{a} = \oint_m (\mathbf{v} \cdot \mathbf{h})\,dl \tag{5.16}$$

and if the number of mean loops per unit volume is ξ, the strain rate is†

$$\dot{\epsilon}_s = b\xi \oint_m (\mathbf{v} \cdot \mathbf{n})\, dl \tag{5.17}$$

In order to show that the concept of mean loops is a useful one, it is necessary to consider the effect of loop shape. This study shows that shape has relatively little effect, so a complex collection of dislocations can be replaced by an equivalent number of mean loops to a good approximation. This is a welcome simplification.

A mean loop might have any shape, but the most representative one that can be readily analyzed is an ellipse. Also, this shape maximizes the ratio of dislocated area to line length (work of formation per unit self-energy), and hence it may be expected that unhampered loops will have elliptical shapes. One such is shown in Fig. 5.5, with j and k its semiaxes parallel and

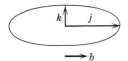

Fig. 5.5 Schematic drawing of a mean elliptical dislocation loop.

perpendicular to its Burgers vector $\mathbf{b}$. The velocities of the edge and screw components of the line at all points are v_e and v_s, so a loop that forms at time $t = 0$ is described by

$$\left(\frac{j}{v_e}\right)^2 + \left(\frac{k}{v_s}\right)^2 = t^2 \tag{5.18}$$

and the loop's area is

$$a = \pi v_e v_s t^2 \tag{5.19}$$

so its glide rate is

$$\dot{a} = 2\pi v_e v_s t \tag{5.20}$$

† If the material does not contain closed loops so an outer normal cannot be clearly defined, then an *equivalent outer normal* must be defined in terms of a unit tangent vector $\mathbf{t}$ at each point. The sign of this is determined by Bilby's right-hand-screw convention, and the equivalent outer normal equals

$$\mathbf{n}_{eq} = \frac{\mathbf{t} \times (\mathbf{b} \times \mathbf{t})}{|\mathbf{b} \times \mathbf{t}|}$$

The extent of the loop in the j direction is $2v_e t$, and twice this extent is the total length of the screw component l_s. Therefore, Eq. (5.20) above may be written

$$\dot{a} = \frac{\pi}{2} l_s v_s$$

or, more generally,

$$\dot{a} = S l_s v_s \qquad (5.21)$$

where S is the *shape factor*. It has a limited range of values, as the following short table of its values indicates:

LOOP TYPE	SHAPE FACTOR
Rectangle	2
Ellipse	$\pi/2$
Straight lines (pair)	1

The limited range of values for S (plus the fact that $\pi/2$ is approximately equal to the root mean square of 1 and 2) means that it is reasonable to replace a complex collection of loops by an equivalent collection of mean elliptic loops with a shape factor of $\pi/2$ and concentration ξ per unit volume. This yields a strain rate of

$$\dot{\epsilon}_s = \frac{\pi}{2} b \xi l_s v_s \qquad (5.22)$$

where the concentration of mean loops is determined by the instantaneous distribution of loop sizes. As this distribution changes through the growth of some loops and the shrinkage of others, so does ξl_s.

5.2 DYNAMICS

In the section on kinematics, some of the consequences of the motions of dislocation lines were considered, but the process by which the dislocations reached steady velocities was not discussed. In order to discuss the acceleration of a dislocation line from rest to some particular velocity, it is necessary to establish an equation of motion. The starting point for this will be Newton's equation in elementary form

$$\sum_1^n f_i = m_d a \qquad (5.23)$$

that is, the sum of the static forces on a small segment of dislocation line must equal the inertial force given by its effective mass m_d times its acceleration. Expressions for various forces and the mass are therefore needed and will be discussed next.

Since a dislocation is a configuration without an intrinsic mass, the forces that act on it are rather special generalized forces. In particular they are unusual because the displacement field of a dislocation is spread out to include the entire volume of a piece of material (if the line is isolated and straight). Therefore, the forces are not localized, but distributed over the configuration.

Several sources of forces may be present. Perhaps the most important is the driving force for gliding motion (the discussion will be restricted to glide; climb is discussed in the literature) (Friedel, 1964).

DRIVING FORCE

In Fig. 5.4a of Sec. 5.1, suppose that a shear stress σ_{12} is present and that the ith dislocation advances a small amount dx_i, thereby causing an external displacement $d\delta_i$. The stress causes an external force

$$F = \sigma_{12}wL$$

which becomes displaced by an amount $d\delta_i = b(dx_i/L)$, with the resulting external work being done

$$dW_{\text{ext}} = F\,d\delta_i = \sigma_{12}bw\,dx_i \tag{5.24}$$

Let the force per unit length that causes the dislocation to move be called $= f$. Then the internal work associated with the dislocation's advance will be

$$dW_{\text{int}} = fw\,dx_i \tag{5.25}$$

By the principle of virtual work, the external work increment must equal the internal one, so

$$f = \sigma_{12}b \tag{5.26}$$

Since the work associated with this force is done when the dislocation moves perpendicular to itself (parallel motions cause no external displacements), the force acts perpendicular to the line.

If more than one stress component is present or if the line is not parallel to one of the coordinate axes, a more general expression for the force is needed. This may be written (Weertman, 1965)

$$\mathbf{F}_g = \mathbf{t} \times \mathbf{G} \tag{5.27}$$

where **t** is the line tangent and **G** is given by

$$G_i = \sum_j^3 \sigma_{ij}^D b_j \qquad (5.28)$$

with b_j = Burgers vector and σ_{ij}^D = deviator stress tensor.

The components of σ_{ij}^D (at the location of the dislocation line) can arise in various ways:

1. Applied stresses.
2. Internal stresses caused by other dislocations or point defects, inclusions, etc.
3. Image stresses near free surfaces.

LINE TENSION

Another important source of force is the line tension which exists because the potential energy of a dislocation line is constant per unit length, so the total energy is proportional to the length. An expression for the line tension can be found by first giving an operational definition of the line tension in terms of the equilibrium of a curved dislocation and then comparing this with a description of the equilibrium of a circular loop in terms of its energy.

A schematic drawing of a curved line with radius r that is in static equilibrium is given in Fig. 5.6. The force $2T$ acts toward the left where T is the

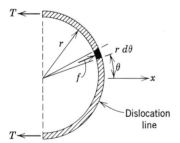

Fig. 5.6 Curved dislocation line with forces acting on it caused by its own line tension and by an applied shear stress on its glide plane.

line tension; and toward the right, a force induced by the applied stress σ_s acts. The stress-induced force on an element of length $r\,d\theta$ is $f = \sigma_s b$, and its component in the x direction is $df_x = f \cos \theta \, r \, d\theta$. The total magnitude of the stress in the $+x$ direction can be found by integration, and equals the tension in the $-x$ direction:

$$2T = r\sigma_s b \int_{-\pi/2}^{+\pi/2} \cos\theta \, d\theta$$

so that

$$T = \sigma_s b r^* \tag{5.29}$$

Another expression for the radius of a dislocation loop that is in equilibrium with an applied shear stress can be found by considering the energy of a circular dislocation loop.

Closed loops are unstable in the absence of an applied stress. Therefore, the energy of a loop in unstable equilibrium with an applied stress can only be found by the method of successive approximations. However, several calculations have been made of arbitrary loop shapes (these have been reviewed by Bacon and Crocker, 1964) including circles, equilateral triangles, rectangles, hexagons, and parallelograms.

For a circular glide loop of radius r, the energy in an isotropic medium is (Kroupa, 1966)

$$U = \frac{G(2 - \nu)}{4(1 - \nu)} b^2 \left(\ln\frac{r}{r_0} - 2 \right) r \tag{5.30}$$

Therefore, the energy is a nonlinear function of the loop radius; but when r is much greater than the size of the core r_0, the energy is nearly proportional to r.

To find the radius that gives unstable equilibrium with an applied stress, a virtual expansion is considered. During such an expansion the virtual work must equal the change in self-energy if the system is in equilibrium. The expression for the virtual work is

$$dw = (\sigma_s b) 2\pi r \, dr \tag{5.31}$$

while the change in self-energy is

$$dU = \left(\frac{2 - \nu}{1 - \nu} \right) \frac{Gb^2}{4} \left(\ln\frac{r}{r_0} - 1 \right) dr$$

Equating the changes yields

$$r^* = \left(\frac{2 - \nu}{1 - \nu} \right) \frac{Gb}{8\pi\sigma_s} \left(\ln\frac{r^*}{r_0} - 1 \right) \tag{5.32}$$

Equating this expression for r^* with that of Eq. (5.29) yields a value for the line tension:

$$T = \left(\frac{2 - \nu}{1 - \nu} \right) \frac{Gb^2}{8\pi} \left(\ln\frac{r}{r_0} - 1 \right) \tag{5.33}$$

This is the tension associated with the elastic self-energy, so it is appropriate that it vanishes when $r = er_0$ where e = base for natural logarithm.

The variation of the line tension with the radius of curvature is

$$\frac{\partial T}{\partial r} = \left(\frac{2 - \nu}{1 - \nu}\right)\frac{Gb^2}{8\pi}\left(\frac{1}{r}\right) \tag{5.34}$$

so there is a strong dependence when r is small, which becomes negligible for large r.

Consider a radius of curvature of $1,000b$ and let $r_0 = b$ with $\nu = \frac{1}{3}$. Then the line tension becomes

$$T \simeq 0.6Gb^2 \tag{5.35}$$

which is about 5 times greater than the surface tension of a solid per atom.

The force that acts tending to straighten a curved dislocation line is

$$f(r) = \frac{T}{r} \tag{5.36}$$

and this is usually small compared with the driving force of an applied stress, unless r is quite small. Therefore, it can often be neglected.

INERTIAL FORCE

Since a dislocation may be considered to be an elastic disturbance, the displacements around it must satisfy the dynamical elastic equilibrium equation (elastic wave equation). The situation is most simple for a screw dislocation because of its cylindrical symmetry which makes two of the displacement components zero, leaving only u_3 for analysis when the line lies parallel to x_3. Then the equilibrium equation becomes

$$G\left(\frac{\partial^2 u_3}{\partial x_1^2} + \frac{\partial^2 u_3}{\partial x_2^2}\right) = \rho\,\frac{\partial^2 u_3}{\partial t^2} \tag{5.37}$$

where the left-hand side gives the net stress force on a small volume element and the right-hand side gives the inertial force. The shear wave velocity is $c = (G/\rho)^{\frac{1}{2}}$.

When the acceleration is zero, static equilibrium prevails, and Eq. (5.37) becomes $\nabla^2 u_3 = 0$; and the solution for a screw dislocation is known to be $u_3 = (b/2\pi)\tan^{-1}(x_2/x_1)$. Since the velocity c is a fixed signal velocity, it is known that the Lorentz transformation of relativity theory will convert the wave equation (5.37) into the Laplace equation. Therefore, the following transformation is substituted into (5.37):

$$x_1^* = \frac{x_1 - vt}{\sqrt{1 - v^2/c^2}} \qquad t^* = \frac{t - vx_1/c^2}{\sqrt{1 - v^2/c^2}}$$

$$x_2^* = x_2 \tag{5.38}$$

$$x_3^* = x_3$$

where x_1^*, t^* refer to the moving coordinate system which moves along the x_1 axis at velocity v.

Let

$$\beta = \sqrt{1 - \frac{v^2}{c^2}}$$

$$x_1^* = \frac{x_1 - vt}{\beta} \qquad \partial x_1^* = \frac{\partial x_1}{\beta}$$

and, if the velocity is steady,

$$\frac{\partial^2}{\partial t^2} = v^2 \frac{\partial^2}{\partial x_1^2}$$

so Eq. (5.37) becomes

$$\frac{\partial^2 x_3^*}{\partial x_1^{*2}} + \frac{\partial^2 u_3^*}{\partial x_2^{*2}} = 0 \tag{5.39}$$

and the appropriate solution is

$$u_3^* = \frac{b}{2\pi} \tan^{-1} \frac{x_2^*}{x_1^*} \tag{5.40}$$

From this displacement field, differentiation yields the strains, and from them the stresses can be obtained by multiplying by the shear stiffness.

The static stress field for the important shear component is

$$\sigma_{\theta 3} = \frac{G}{r}\left(\frac{\partial u_3}{\partial \theta}\right) = \frac{Gb}{2\pi r}$$

which may be compared with the dynamical case

$$\sigma_{\theta 3}^* = \frac{Gb}{2\pi r}\left(\frac{\beta}{\cos^2 \theta + \beta^2 \sin^2 \theta}\right) \tag{5.41}$$

For the moving dislocation the stress on the glide plane $\theta = 0°$ is $Gb\beta/2\pi x_1$, which means that a given magnitude lies closer to the dislocation's center than in the static case. On the other hand, for $\theta = 90°$ the shear stress is $\sigma_{\theta 3}^* \rightarrow Gb/2\pi r\beta$, so a given magnitude lies further away from the center. Figure 5.7 shows a sketch of the stress trajectory as it compares with the

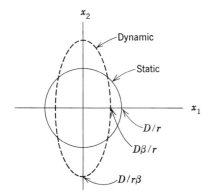

Fig. 5.7 Change in shape of stress contours when a dislocation is set into motion.

static case. It may be seen that the stress field becomes contracted during motion.

The strain energy of a static screw dislocation is given by

$$U_0 = \frac{Gb^2}{4\pi} \ln \frac{R}{r_0} \tag{5.42}$$

and for an observer moving with the moving coordinate system, this is unchanged. However, for a stationary observer, the potential energy becomes

$$U = U_0 \frac{1 + \beta^2}{2\beta} \tag{5.43}$$

while the kinetic energy is

$$T = \int\int \frac{1}{2} \rho \left(\frac{\partial u_3^*}{\partial t}\right)^2 2\pi r \, dr \, d\theta \tag{5.44}$$

$$T = U_0 \left(\frac{v}{c}\right)^2 \frac{1}{2\beta} \tag{5.45}$$

so the total energy is

$$U_T = U + T = \frac{U_0}{\beta} \tag{5.46}$$

which becomes very large as v approaches c, making $\beta \rightarrow 0$.

For very low velocities, $\beta \to 1$, so Eq. (5.45) may take the form

$$T = \frac{1}{2} \left[\frac{U_0}{c^2} \right] v^2 \tag{5.47}$$

where the term in brackets is the analog of the mass in the theory of particles. Thus an *effective rest mass* m_0 may be described

$$m_0 = \frac{U_0}{c^2} = \frac{\rho b^2}{4\pi} \ln \frac{R}{r_0} \tag{5.48}$$

For a typical specimen, $\ln R/r_0 \simeq 4\pi$ so $m_0 \simeq \rho b^2$ or an atomic mass per unit length.

It is also useful to notice that $U_0 = m_0 c^2$, which is an analog of Einstein's equation for particles. If this analog is carried further, the energy of a moving dislocation is

$$U = mc^2 \tag{5.49}$$

so the dynamic mass is

$$m = \frac{m_0}{\beta} \tag{5.50}$$

and, as $v \to c$, the effective mass increases without limit.

Analogous expressions are obtained for edge dislocations (see Weertman, 1961, for a review).

Since a dislocation line has an effective mass because its displacement field moves matter when it moves, an inertial force appears whenever it is accelerating, and this force opposes the driving force.

VISCOUS DRAG FORCE

Moving dislocations dissipate most of the work that is done on them in the form of heat. Less than 5 percent is stored in the material as internal energy. The mechanisms by which this occurs will be considered elsewhere, but it requires that viscous forces exist which represent the dissipation process.

In general, the viscous forces depend in a nonlinear way on the dislocation velocity. However, for the present purposes the most simple case will be considered, that of linear (newtonian) viscosity. Then, if B is the viscous damping coefficient, the viscous force per unit length is

$$f_v = Bv \tag{5.51}$$

where v is the dislocation velocity.

EQUATION OF MOTION AND TRANSIENT BEHAVIOR

A straight dislocation line in dynamical equilibrium has three forces acting on it. One tends to move it forward; this is the driving force $f_d = \sigma_s b$ caused by stress. Two tend to oppose its motion; these are the inertial force $f_i = ma$ and the viscous drag force $f_v = Bv$. Thus,

$$f_d = f_i + f_v \tag{5.52}$$

or

$$\sigma_s b = m \frac{dv}{dt} + Bv \tag{5.53}$$

Upon separating the variables and integrating, this yields

$$v(t) = \frac{\sigma_s b}{B} \left(1 - e^{-(B/m)t}\right) \tag{5.54}$$

so the velocity is zero at $t = 0$ when the stress is applied. It then rises at a rate that is determined by the acceleration coefficient (time constant) B/m, that is, by the ratio of the viscosity and inertia coefficients. As the velocity increases, the viscous force approaches the magnitude of the driving force, so the acceleration becomes small and a steady-state velocity v_{ss} is approached:

$$v_{ss} = \frac{\sigma_s b}{B} \tag{5.55}$$

This steady-state velocity is determined by the ratio of the driving force to the viscous drag coefficient.

For a pure metal, B might be as small as 10^{-3} P. Then for a stress of 10^8 dyn/cm² (100 atm), and letting $b = 3A$, the steady-state velocity would be 3×10^3 cm/sec; and if the effective mass is taken to be ρb^2 with $\rho = 8$ g/cm³, this velocity would be reached in about 7×10^{-12} sec. This rapid transient is very short compared to ordinary fast loading times of about 1 μsec, and even most shock fronts do not load solids as fast as this. Note that larger viscosity coefficients produce even shorter transients, but with lower steady-state velocities.

5.3 STEADY-STATE MOTION

Because of the rapid acceleration of dislocations in most circumstances, the period of transient motion is very short, and most of the time is spent in steady-state motion. Two qualifications must be put on this statement.

First, it should be noted that *transient* here refers to individual line segments. If some form of dislocation multiplication occurs, then there can be a relatively long period of transient plastic flow in a specimen taken as a whole. Second, the term *steady-state* refers to motion that is averaged over a substantial time period. Since many forces act on moving dislocations, their velocities constantly oscillate about the mean steady-state value. Even in a perfect crystal, the discrete atomic structure causes small oscillations; and in imperfect ones, large excursions from the mean can be caused by interactions with defects.

Equation (5.55) of Sec. 5.2 shows that the steady-state velocity is determined by the viscosity coefficient B. The role played by this quantity can be examined further by calculating the power-loss field around a moving dislocation. This was first done by Mason (1960). An improved method will be used here (Gilman, 1968).

THE DRAG ON A MOVING DISLOCATION IN A VISCOUS MEDIUM

Suppose that a straight dislocation line lies parallel to the x_3 axis and moves along a glide plane in the x_3 direction. Then a velocity gradient $\partial v_1/\partial x_2$ exists in the x_2 direction (that is, perpendicular to the glide plane). If momentum can be transported from the higher to the lower velocity regions, viscous loss occurs whose magnitude is determined by the viscosity coefficient defined by

$$\sigma_s = \eta \frac{\partial v_1}{\partial x_2} \tag{5.56}$$

and assumed to be constant. Also, it may be noted that $\partial v_1/\partial x_2 = \dot{\epsilon}_{12}$, the shear strain rate.

The power loss dP in a small volume element dV depends on the square of the strain rate within it and on η:

$$dP = \eta(\dot{\epsilon}_{12})^2 \, dV \tag{5.57}$$

To calculate the total loss, the strain-rate field may be divided into two parts, one immediately at the glide plane and the other outside the glide-plane region. If the separation distance at the glide plane is a, then in the regions $x_2 < -a/2$ and $x_2 > +a/2$ the strain rate can be obtained to a good approximation from the elastic strain field of a dislocation. For a screw dislocation, the rate field is

$$\dot{\epsilon}_{12} = \frac{b}{2\pi} \frac{v_1 \cos \theta}{r^2} \tag{5.58}$$

This can be substituted into Eq. (5.57) and integrated:

$$P_1 = 2\eta \left(\frac{bv}{2\pi}\right)^2 \int_{a/2}^{\infty} \int_{-\infty}^{+\infty} \frac{x_1^2}{(x_1^2 + x_2^2)^3}\, dx_1\, dx_2 = \frac{\eta b^2 v^2}{8\pi a^2} \tag{5.59}$$

In the region $-a/2 < x_2 < +a/2$, the relative displacement across the glide plane is given to a good approximation by

$$u(x_1) = -\frac{b}{\pi} \tan^{-1} \frac{2x_1}{w} \tag{5.60}$$

where w is the "width" of the dislocation defined as in Fig. 5.2 of Sec. 5.1.

The expression for the velocity gradient at the glide plane may be written

$$\dot{\epsilon}_0 = \frac{1}{a}\left(\frac{du}{dt}\right) = \frac{v_1}{a}\left(\frac{\partial u}{\partial x_1}\right) \tag{5.61}$$

This also may be substituted into Eq. (5.57) and integrated:

$$\begin{aligned} P_0 &= \frac{4\eta v^2 b^2}{\pi^2 w^2 a} \int_{-\infty}^{+\infty} \left[1 + \left(\frac{2x_1}{w}\right)^2\right]^{-2} dx_1 \\ &= \frac{\eta v^2 b^2}{\pi a w} \end{aligned} \tag{5.62}$$

Equations (5.59) and (5.62) may be combined to obtain the total power loss:

$$P_t = \frac{\alpha \eta v^2 b^2}{\pi a w}\left(1 + \frac{1}{8}\frac{w}{a}\right) \tag{5.63}$$

The factor α is unity for a screw and $3/4(1 - \nu)$ for an edge dislocation. The width is usually comparable with a in size, so the second term in (5.63) is small compared with the first, and the drag force per unit length may be written

$$F_v = \frac{P_t}{v} \simeq \frac{\eta v b^2}{\pi a w} \tag{5.64}$$

This implies that low viscous drag is favored by a large width and glide-plane spacing and by a small Burgers displacement. All these factors are favorable for extended dislocations that lie on close-packed planes; this accounts in part for their high mobilities.

SOURCES OF VISCOUS DRAG

According to the kinetic theory of liquids (Green, 1952), there are two general categories of viscous mechanisms: these are "*gaslike*" and "*solidlike.*" Both tend to smooth out differences in velocity between different parts of a fluid. In the gaslike mode, particles (or quasi-particles) are free to traverse relatively long distances between collisions. Then, if two adjacent layers of a medium are moving at different velocities, the average particle velocities in the two layers must differ. Therefore, particles that move from the faster layer to the slower have more velocity on the average than the particles that are already in the slower layer. Thus the velocity of the slower layer tends to increase, while that of the faster tends to decrease, and velocity differences tend to diminish. Furthermore, somewhere in the system there must be a reference layer with zero velocity (for an observer in the fluid). Since momentum cannot be transferred to the reference layer, the whole fluid must tend to lose velocity and come to rest in the absence of a net driving force.

In the solidlike mode, it is direct intermolecular interactions that tend to smooth velocity differences. The molecules are constrained to remain in their own layers, but a faster-moving layer sliding over a slower one exerts a dragging force that tends to speed up the latter. At the same time the faster layer tends to slow down.

For dislocations, both of the above modes can operate at the core where sliding occurs, and the gaslike mode can operate in the surrounding elastic strain-rate field. Several detailed mechanisms participate in causing viscous losses, and they will be discussed further in Chap. 8.

EXPERIMENTAL DISLOCATION-MOBILITY MEASUREMENTS

There are two general methods of measuring dislocation mobilities. One uses direct observations of positions by means of selective etching (Gilman and Johnston, 1957), while the other uses indirect analysis of ultrasonic attenuation at very high frequencies (Stern and Granato, 1962). The former allows average velocities to be determined for relatively large displacements ($1 \ \mu$ and up), while the latter allows a damping coefficient to be determined for small displacements.

Selective etching before and after a stress pulse has been applied to a crystal determines the initial and final positions of dislocations and thus the distance moved Δx. The motion occurs within the duration of the stress pulse Δt, since substantial damping is present and since the time spent for acceleration is very short. Thus the average velocity is $\bar{v} = \Delta x / \Delta t$. Figure 5.8 illustrates an actual observation.

Several means for producing stress pulses have been used, depending on

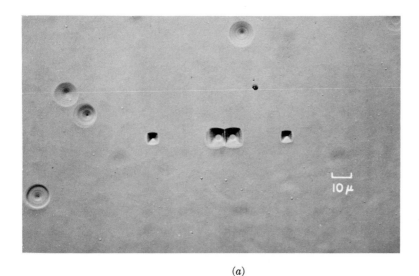

(a)

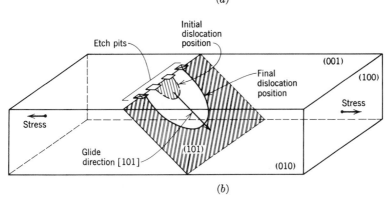

(b)

Fig. 5.8   Etch pits showing motion of an individual dislocation loop in a lithium fluoride crystal. (a) Photograph; (b) schematic drawing.

the time interval of interest:

TIME INTERVAL	METHOD
>5 sec	Dead weight
>1 sec	Lever with rolling weight
>10^{-2} sec	Electromagnets
>2×10^{-6} sec	{ Rebound of elastic spheres Torsion shock bar

Velocities have been measured in a variety of substances, and some representative results are shown in Fig. 5.9.

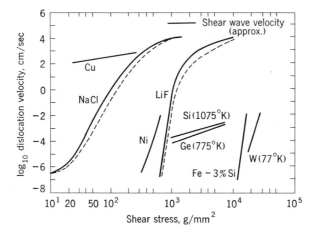

Fig. 5.9 Some typical data showing the dependence of dislocation velocities on applied shear stress.

The data summarized by Fig. 5.9 have several distinctive features. One is that the velocities are spread over a very large range (nearly 12 orders of magnitude for LiF). Another is that for some substances, quite small stress changes cause very large velocity changes until very high velocities are reached, where the velocity increases slowly with further stress increases. Edge dislocations move faster at a given stress level than screw dislocations, most likely because the latter tend to be jogged and the jogs leave trails of edge dipoles. The dipoles require energy for their formation, which causes a drag force on the moving screw dislocation (note that this is different from a viscous drag, but has a similar mechanical effect).

It is apparent from Fig. 5.9 that no general relation between dislocation velocity and stress can be written. In some cases, such as those of Cu and Ge, a linear relation is observed; but in others, such as LiF and NaCl, the relation is very nonlinear. There are some limiting cases, however, according to the type of chemical binding in the crystal and whether the motion is induced by stress alone or is assisted by thermal fluctuations. One way of classifying the cases is as follows:

1. Local Bonding (covalent):
    a. Low stress levels ($\gtrsim G/100$); motion thermally activated, velocity vanishes at low temperatures.
    b. High stress levels ($\lesssim G/100$); stress alone causes motion at low temperatures.

2. Nonlocal Bonding (pure simple metals and salts):
   a. *Low stress levels;* motion occurs at lowest temperatures and stresses.
      (1) Salts; phonons cause drag at high temperatures.
      (2) Metals; electrons cause drag at low temperatures, phonons at high.
   b. *High stress levels;* viscosity increases because of relativistic effects.

In typical real metal and salt crystals, localized bonding occurs near impurities and other defects. This creates a heterogeneous system whose behavior may be complex.

For any type of crystal, the dislocation velocity cannot increase without limit as the stress increases, because loss mechanisms are present in all real crystals. It must saturate below the velocity of sound waves if the stress field is continuous (supersonic motion is possible for discontinuous stress fields). If the limiting velocity is called v^*, the velocity expression may be written quite generally as

$$v = v^*P_m(\sigma_s) \tag{5.65}$$

where P_m is the average probability for the dislocation to have the velocity v^* at a given instant. This probability is a function of the stress and may also depend on the prior plastic strain, the temperature, impurity concentration, etc.

A functional form for P_m must have a zero value when $\sigma_s = 0$ and must approach unity asymptotically for large σ_s. One function that fits much of the data and has these limits is (Gilman, 1960)

$$e^{-D/\sigma_s} \tag{5.66}$$

where D is called the characteristic drag stress. This function provides a good representation of data at medium to high stress levels. Also, it has a sudden transition from low to high values when $\sigma_s \simeq D/2$, so it represents plastic "yielding" quite well. However, its form at low stress levels is poor, because its derivatives are complicated. Also, it cannot represent the linear velocity-stress relations that are observed for Cu, Ge, and Zn. It can be derived from an atomic model of "stress activation."

Linear behavior at small stress levels plus saturation at high ones can be represented by (see also Chap. 8)

$$1 - e^{-\sigma_s/S} \tag{5.67}$$

where S is the coupling stress that acts across the glide plane. For small stresses this becomes σ_s/S, so the velocity is proportional to the stress, and the drag coefficient is

$$B = \frac{bS}{v^*} \tag{5.68}$$

The behavior of almost any material can probably be represented by a combination of Eqs. (5.65), (5.66), and (5.67):

$$v = v_s^*(1 - e^{-\sigma_s/S}) + v_d^* e^{-D/\sigma_s} \tag{5.69}$$

This procedure is arbitrary, but consistent with the fact that real materials are heterogeneous, so that more than one mechanism controls their behavior.

Information about dislocation mobilities can also be obtained from an interpretive analysis of ultrasonic attenuation measurements. The analysis begins with the equation of motion for a dislocation line that has a length L and vibrates like a stretched string (Koehler, 1952). Let the line lie parallel to the x axis and move in the xy plane when it is acted upon by a shear stress σ_{zy}. The driving force on it is then $b\sigma_{zy} = b\sigma_0 \sin \omega t$ if a cyclic stress is applied with angular frequency ω. A drag force $Bv = B(\partial y/\partial t)$ opposes the driving force, as does an inertial force $m(\partial^2 y/\partial t^2)$, whereas the line-tension force $T(\partial^2 y/\partial x^2)$ aids the applied force. Here B is a damping constant, m is the effective mass per unit length, and T is the line tension.

In the absence of damping ($B = 0$), and for small amplitudes of vibration, resonance occurs at a frequency

$$\omega_0 = \frac{\pi}{L}\sqrt{\frac{T}{m}} \tag{5.70}$$

For large amounts of damping, the attenuation reaches a maximum at a frequency (Granato and Lucke, 1956)

$$\omega = \frac{\pi^2 T}{BL} \tag{5.71}$$

At frequencies that are higher than this (typically in the range 10–100 MHz), the dislocations no longer vibrate like strings, but are more like rigid rods, and the ultrasonic attenuation reaches a limiting value

$$\alpha = \frac{8\phi TN}{\pi^2 B} \tag{5.72}$$

where ϕ is an orientation factor that relates xyz to the coordinates of the sound wave and N is the dislocation density. Thus, if N and T are known, B can be measured (Granato, 1968). Some numerical values are listed in Table 5.1.

TABLE 5.1 DAMPING CONSTANTS FOR DISLOCATION MOTIONS

CRYSTAL	DAMPING CONSTANT $= B(10^{-4}\text{ dyn-sec/cm}^2)$			
	FROM VHF ULTRASONIC ATTENUATION	REF.	FROM SLOPE OF LINEAR VELOCITY-STRESS CURVE	REF.
LiF	16.0	b	7.0	a
	2.5	c		
	2.4	e		
NaCl	(2.5–10.5)	d		
	1.6	e		
KCl	3.2	b		
KBr	—	—	~20	f
Al	~10.0	l		
Cu	8.0	g	7.0	i
	6.5	h		
	1.2	b	1.4	m
Zn	—	—	7.8	j
Pb	~3.7 (300°K)	k		
	~1.1 (60°K)			

a. W. P. Mason, *J. Acoust. Soc. Am.*, **32**: 458 (1960).
b. T. Suzuki, A. Ikushima, and M. Aoki, *Acta Met.*, **12**: 1231 (1964).
c. O. M. Mitchell, *J. Appl. Phys.*, **36**: 2083 (1965).
d. R. A. Moog, Ph.D. thesis, Cornell University, 1965.
e. F. Fanti, J. Holder, and A. Granato, *J. Am. Acoust. Soc.*, June (1969).
f. V. B. Pariiskii, S. V. Lubenets, and V. I. Startsev, *Sov. Phys.-Solid State*, **8**: 976 (1966).
g. G. A. Alers and D. O. Thompson, *J. Appl. Phys.*, **32**: 283 (1961).
h. R. M. Stern and A. V. Granato, *Acta Met.*, **10**: 358 (1962).
i. W. F. Greenman, T. Vreeland, and D. S. Wood, *J. Appl. Phys.*, **38**: 3595 (1967)
j. D. P. Pope, T. Vreeland, and D. S. Wood, *J. Appl. Phys.*, **38**: 1929 (1967).
k. W. P. Mason and A. Rosenberg, *J. Appl. Phys.*, **38**: 1929 (1967).
l. ——— and ———, *Phys. Rev.*, **151**: 434 (1966).
m. T. Suzuki, "Dislocation Dynamics," McGraw-Hill Book Company, New York, 1968.

Values of B may also be obtained for velocity-stress curves, and some of these are listed in Table 5.1 for comparison with the ultrasonic values. The two sets of values agree approximately for Cu and LiF. If the tabulated values for B are compared with the viscosities of liquid metals, it is found that they are typically 10 to 100 times smaller than liquid-metal viscosities at the melting point.

THERMAL STABILITY

Since a moving dislocation converts most of the work done on it into heat, the possibility that the heat from one part of a line acts on other parts to influence their behavior must be considered. Since the flow stress is sometimes a strong function of temperature, local heating might lower it enough to cause easy local propagation and self-sustained instability. Such instability is well-known on a macroscopic scale during high-speed deformation (Zener, 1948).

Eshelby and Pratt (1956) have studied the temperature distribution around a moving dislocation line with Burgers displacement b when the line is acting as a heat source of strength:

$$P = bv\sigma_s \tag{5.73}$$

If the line lies parallel to the x_2 axis and moves with velocity v in the x_1 direction under an applied shear stress σ_s, then the temperature at a point (x_1, x_2) relative to the center of the dislocation is

$$T = \frac{P}{2\pi h} e^{x_1/\Lambda} K_0 \left(\frac{r}{\Lambda}\right) \tag{5.74}$$

where $K_0(Z)$ = modified Bessel function
$\quad r^2 = x_1^2 + x_2^2$
$\quad \Lambda = 2H/V$ = characteristic distance
$\quad H$ = thermal diffusivity = h/C_v
$\quad h$ = thermal conductivity
$\quad C_v$ = specific heat at constant volume

The "size" of the temperature-distribution pattern is measured by Λ, and some typical values are as follows:

MATERIAL	H (cm²/sec)	V_s (cm/sec) (sound velocity)	$\lambda_{min}(A) = 2H/V_s$
Ag	1.70	1.01	2100
Al	0.86	3.04	570
Fe	0.17	3.24	100
glass	0.0057	3.3	3.5
Al₂O₃	~0.025	5.1	~9.8
SiC	~0.057	5.5	~21.0

The asymptotic values of the modified Bessel function are

$$K_0(Z) \to -\ln Z \qquad (Z \ll 1)$$
$$\to \left(\frac{\pi}{2Z}\right)^{\frac{1}{2}} e^{-Z} \qquad (Z \gg 1) \tag{5.75}$$

Then if $\Lambda \gg b$, the temperature near the dislocation is

$$T \simeq \frac{b\sigma_s}{2\pi h} \left(v \ln \frac{2H}{vr} \right) \qquad (5.76)$$

so the variation of the temperature with velocity is

$$\frac{\partial T}{\partial v} \simeq \frac{b\sigma_s}{2\pi h} \left(\ln \frac{\Lambda}{r} - 1 \right) \qquad (5.77)$$

For at least a narrow temperature range the steady-state dislocation velocity can be expressed as

$$v = v^* e^{-B/\sigma_s T} \qquad (5.78)$$

Then, if the motion is steady, the variation of the velocity with temperature is

$$\frac{\partial v}{\partial T} = \frac{Bv}{\sigma_s T^2} \qquad (5.79)$$

Therefore, if the rate of temperature rise given by Eq. (5.77) exceeds the inverse of Eq. (5.79), instability can be expected, because v will tend to increase at constant stress; that is, instability may occur for

$$\frac{b\sigma_s}{2\pi h} \left(\ln \frac{\Lambda}{r} - 1 \right) > \frac{\sigma_s T^2}{Bv} \qquad (5.80)$$

but $B = D/T$ where D is the characteristic drag stress, so this may be substituted, and Eq. (5.80) may be solved for the critical velocity:

$$v_{\text{crit}} > \frac{2\pi T h}{Db} \left(\ln \frac{\Lambda}{r} - 1 \right)^{-1} \qquad (5.81)$$

Consider two prototype materials, silver and aluminum oxide, at room temperature (300°K) and let $b = 3A = r$ for both of them:

MATERIAL	$h \left(\dfrac{\text{ergs}}{\text{sec-cm-°C}} \right)$	D (dyn/cm²)	(Λ/r) for $v = v_s/10$	$\sim v_{\text{crit}}$ (cm/sec)
Ag	4.2×10^7	$G/10^3$	$\sim 7 \times 10^3$	10^9
Al₂O₃	0.34×10^7	$G/10$	~ 15	7×10^4

This leads to the conclusion that metals tend to be stable at all velocities, but insulators may show dynamic softening. At low temperatures where the specific heat becomes small, even metals may become unstable (Erdmann and Jahoda, 1968).

REFERENCES

Bacon, D. J., and A. G. Crocker: The Elastic Energies of Dislocation Loops, *International Conference on Lattice Defects in Quenched Metals*, Argonne National Laboratory, 1964.

Erdmann, J. C., and J. A. Jahoda: *J. Appl. Phys.*, **39**: 2793 (1968).

Eshelby, J. D., and P. L. Pratt: *Acta Met.*, **4**: 560 (1956).

Friedel, J.: "Dislocations," p. 104, Pergamon Press, New York, 1964.

Gilman, J. J.: *Australian J. Phys.*, **13**: 327 (1960).

———: *Phys. Rev. Letters*, **20**: 157 (1968).

——— and W. G. Johnston: "Dislocations and Mechanical Properties of Crystals," p. 116, John Wiley & Sons, Inc., New York, 1957.

Granato, A. V.: in A. R. Rosenfield et al. (eds.), "Dislocation Dynamics," p. 117, McGraw-Hill Book Company, (1968).

——— and K. Lucke: *J. Appl. Phys.*, **27**: 583 (1956).

Green, H. S.: "The Molecular Theory of Fluids," Interscience Publishers, New York, 1952.

Koehler, J. S.: in Shockley et al. (eds.), "Imperfections in Nearly Perfect Crystals," p. 197, John Wiley & Sons, Inc., New York, 1952.

Kroupa, F.: Dislocation Loops, in B. Gruber (ed.), "Theory of Crystal Defects," p. 275, Academic Press Inc., New York, 1966.

Laub, T., and J. D. Eshelby: *Phil. Mag.*, **14**: 1285 (1966).

Mason, W. P.: *J. Acoust. Soc. Am.*, **32**: 458 (1960).

Seeger, A., and P. Schiller: *Acta Met.*, **10**: 348 (1962).

Stern, R. M., and A. V. Granato: *Acta Met.*, **10**: 358 (1962).

Weertman, J.: *Phil. Mag.*, **11**: 1217 (1965).

———: High Velocity Dislocations in Shewman and Zackay (eds.), "Response of Metals to High Velocity Deformation," p. 205, Interscience Publishers, New York, 1961.

Zener, C.: The Micromechanism of Fracture in "Fracturing of Metals," p. 3, American Society for Metals, Cleveland, Ohio, 1948.

6

POPULATION OF
MOBILE DISLOCATIONS

During plastic deformation, the internal structure of a crystal changes. The total dislocation density increases as a result of regenerative processes, and the increase tends to facilitate further flow. On the other hand, further flow tends to be inhibited by interactions between dislocations, which become increasingly frequent as the population density increases. Also, after large strains have taken place, whole blocks of material within crystals often become rotated with respect to the remainder (the kink bands of Fig. 3.16 are a simple form of this). That is, a crystal tends to become "fragmented" so its glide planes are no longer flat. This complicates the ways in which dislocations interact.

The behavior that is outlined above is too complex to be described in detail, and since one or two processes do not dominate the situation, it is this author's view that only the average behavior can be described. This is best done in terms of a few measurable parameters with little or no attempt to develop a detailed theory. The situation is somewhat analogous to the turbulent flow of fluids where detailed processes can be studied (and are), but the overall macroscopic behavior of a fluid is described in terms of a few measurable properties. It is nevertheless important to be aware of the complex internal structure in order to avoid pitfalls when the theory is applied to unfamiliar situations.

The effect of interactions on the mobile population can be described in either of two ways from a statistical viewpoint. One is to consider that the average dislocation velocity decreases as a result of plastic strain, because collisions tend to reduce the drift velocity. The other is to consider that the moving dislocations continue to move at the same speed, but the fraction of the total density that moves decreases continually as the plastic strain increases. The two viewpoints are statistically equivalent.

There are three basic cases that require description:

1. Small strains where multiplication dominates
   a. Fixed source concentration
   b. Breeding via multiple cross glide
   c. Spontaneous nucleation
2. Approach to steady state at large strains
3. Hardening caused by immobilization at large strains

Since a statistical description is sought, the behavior of a single source or line is not considered. Instead, attention is concentrated on an assemblage of random elemental events.

6.1 SMALL-STRAIN CASE

FIXED CONCENTRATION OF SOURCES

In order to simplify the discussion, the sources are taken to be small compared with the specimen size (point sources). Then, in a thin slab of thickness w and unit area, there will be wS sources if S is the number of sources per unit volume. If each source emits a pair of dislocation lines at a rate r per second, the rate of change of dislocation lines per unit area in the slab will be

$$\dot{N}_s = rwS \tag{6.1}$$

and the rate of change in a thick slab will be proportional to this. Thus $\dot{N}$ is constant, and N increases in proportion to time:

$$N = N_0 + Kt \tag{6.2}$$

Substituting into the equation for the strain rate,

$$\dot{\epsilon} = bv(N_0 + Kt) \tag{6.3}$$

and neglecting N_0 compared with Kt, a relation between N and the strain may be found by integration:

$$\epsilon = \frac{1}{2}\frac{bvN^2}{K}$$

If this is solved for N,

$$N = A\sqrt{\epsilon_p} \tag{6.4}$$

where A is the constant $(2K/bv)^{\frac{1}{2}}$, so the dislocation density is proportional to the square root of the plastic strain. This is rarely observed experimentally.

BREEDING VIA MULTIPLE CROSS GLIDE

If a crystal is not too anisotropic, it is possible for screw dislocations to cross glide and therefore breed (Koehler, 1952; Johnston and Gilman, 1960). The geometry of this process was described in Sec. 4.7. In effect, new sources of dislocations are created during the flow process itself, so the rate of change of N is proportional to N (Fig. 6.1). Furthermore, it is experimentally

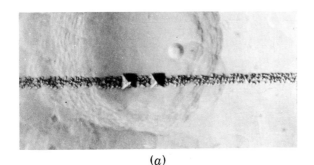

(a)

(b)

Fig. 6.1 Breeding of dislocations in LiF: (a) glide band containing many dislocations that formed when stress was applied to a single half loop whose ends were located at the centers of the two large etch pits (note that the small pits show that new dislocations lie on both sides of the glide plane of the original loop); (b) widening of a glide band. Crystal was deformed twice and etched after each deformation. W_1 and W_2 show the glide band widths after the two deformations. Magnification = 500×.

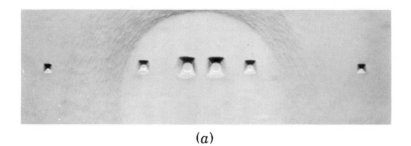

(a)

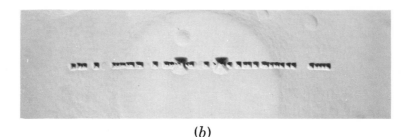

(b)

Fig. 6.2 Multiplication of dislocation half loops: (a) half loop expanded slowly in two stages (no visible multiplication); (b) half loop expanded more rapidly with 30 percent greater stress (profuse multiplication).

observed (Johnston and Gilman, 1959) that the multiplication rate (increase in total loop length per unit distance moved by the initial line) increases with the mean velocity (Fig. 6.2). This can also be justified theoretically (Li, 1961). Therefore, if the breeding coefficient is called m, the first-order kinetic equation becomes

$$\dot{N} = mvN \qquad (6.5)$$

and, upon integration,

$$N = N_0 e^{vmt} \qquad (6.6)$$

which gives a rapid increase of density with time. In particular, the logarithm of N should increase in proportion to the time, and this has been verified experimentally for LiF crystals.

An important consequence of the breeding behavior expressed in Eq. (6.5) is that the dislocation density increases in proportion to the total plastic strain. This can be deduced by substituting Eq. (6.5) into the strain-rate equation to obtain

$$\dot{\epsilon} = \frac{bv\dot{N}}{mv} = \frac{b}{m}\,\dot{N} \tag{6.7}$$

this may then be multiplied by dt and integrated:

$$N(\epsilon_p) = N_0 + \frac{m}{b}\,\epsilon_p \tag{6.8}$$

A linear dependence of N on ϵ_p has been observed by numerous investigators, and approximate values for the coefficient $M = m/b$ are listed in Table 6.1. Note that if Eq. (6.8) results from experiment, then Eq. (6.5) is a required consequence.

Since m is determined by the relative rate of cross glide, it is expected to vary with crystal anisotropy, with the resolved shear stress on the cross-glide plane, and with temperature. Rather little information is available concerning these dependences, however, so M must be treated as a measurable parameter.

It should be emphasized here that Eq. (6.8) has considerable significance because it simplifies the theory of plastic flow markedly. This makes the theory more than simply a restatement of known facts in symbolic form. The simple linear form of the equation results from the first-order kinetics of Eq. (6.5). Also, this latter equation contains the dislocation velocity, and this changes with stress, so the multiplication coefficient m is independent of stress to a first approximation (Gilbert, Wilcox, and Hahn, 1965). Thus some rather complicated behavior is summarized very compactly in Eq. (6.8). Nevertheless, it must be applied with some caution because covalent crystals do not seem to obey it (Alexander and Haasen, 1968).

SPONTANEOUS NUCLEATION

It has long been recognized that the large self-energies of dislocation lines make it very difficult to form them in a perfect crystal (Seitz and Read, 1941). Theoretical estimates place the stress required for homogeneous nucleation at about $\frac{1}{30}$ of the shear modulus of a crystal (Cottrell, 1953, p. 53). This has been verified in filamentary iron crystals by Brenner (1956) and in massive LiF crystals by Gilman (1959). Plastic deformation usually occurs at stresses that are 100 to 1,000 times smaller than the above value, so homogeneous nucleation is not expected to contribute to plasticity under ordinary conditions. It might occur at substantial rates during shock loading, however.

Several situations can lead to heterogeneous nucleation, as shown by

TABLE 6.1 DISLOCATION MULTIPLICATION COEFFICIENTS (All values approximate)

MATERIAL	$M(10^9/cm^2)$	REFERENCE
Ag	230	J. E. Bailey and P. B. Hirsch, *Phil. Mag.*, **5**: 485 (1960).
Al	80	M. J. Hordon and B. L. Averbach, *Acta Met.*, **9**: 247 (1961).
Cu (monocrystal)	100	M. J. Hordon and B. L. Averbach, *Acta Met.*, **9**: 247 (1961).
Cu (monocrystal)	2	J. D. Livingston, *Acta Met.*, **10**: 229 (1962).
Cu (polycrystal)	50	J. E. Bailey, reported by Livingston above.
Fe	33	W. E. Carrington, K. F. Dale, and D. McLean, *Proc. Roy. Soc. (London)*, **259A**: 203 (1960).
Fe (coarse grain, 100 μ)	80	A. S. Keh and S. Weissmann, "Electron Microscopy and Strength of Crystals", Interscience Publishers, New York, 1963.
Fe (fine grain, 15 μ)	200	
Fe	250	G. T. Hahn, C. N. Reid, and A. Gilbert, *Phil. Mag.*, **11**: 409 (1965).
Fe (3% Si)	200	J. R. Low and A. M. Turkalo, *Acta Met.*, **10**: 215 (1962).
Ge	1	J. R. Patel and B. H. Alexander, *Acta Met.*, **4**: 385 (1956).
	0.7	R. L. Bell and W. Bonfield, *Phil. Mag.*, **9**: 9 (1964).
InSb	1	M. S. Abrahams and W. K. Liebmann, *Acta Met.*, **10**: 941 (1962).
LiF	1	J. J. Gilman and W. G. Johnston, *J. Appl. Phys.*, **31**: 687 (1960).
Ni	8	W. M. Yim and N. J. Grant, *Trans. AIME*, **227**: 868 (1963).
Mo	80	G. T. Hahn and A. Gilbert, *Phil. Mag.*, **11**: 409 (1965).
	90	A. Lawley and H. L. Gaigher, *Phil. Mag.*, **10**: 15 (1964).
Ta	1,000	W. S. Owen, *Phil. Mag.*, **11**: 409 (1965).
U	400	D. L. Douglass and S. E. Bronisz, *Trans. AIME*, **227**: 1151 (1963).
V	500	J. W. Edington and R. E. Smallman, *Acta Met.*, **12**: 1313 (1964).

Gilman (1959) and others. Usually some form of discontinuity creates a stress concentration large enough to produce dislocations. These include (see Figs. 6.3, 6.4, and 6.5):

1. Surface steps produced by:
    a. Growth
    b. Cleavage

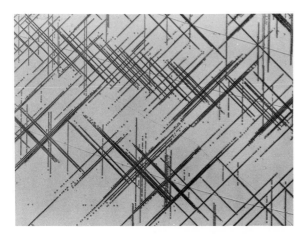

Fig. 6.3 Glide band that formed at cleavage steps when a stress pulse was applied to As-cleaved LiF. The lines that run from upper left to lower right are the cleavage steps. Magnification = 100×.

2. Inclusions:
   a. Impurity precipitates ∼0.1 μ diameter and larger
   b. Clusters induced by radiation damage
   c. High-pressure gas bubbles (Clarebrough et al., 1967)
3. Glide bands: these sometimes spawn secondary bands that spread transversely from the primary ones
4. Subgrain boundaries

The kinetic behavior of these heterogeneous sources is similar to that of regenerative ones such as the Frank-Read source. Therefore the description given in Sec. 6.1 applies.

6.2 LARGER-STRAIN BEHAVIOR

As plastic straining proceeds and the population of dislocations increases, various kinds of interaction occur with increasing frequency. Parallel dislocations interact elastically to form dipoles, tripoles, and higher-order multipoles (Sec. 4.8). Nonparallel dislocations cut through one another and then trail dipoles behind them (Sec. 4.6). Dislocations interact inelastically to form new hybrids (Sec. 4.14) or to annihilate one another (Sec. 4.14). Some dislocations move into grain boundaries or out of free surface and become lost. There are so many possibilities that only an overall statistical

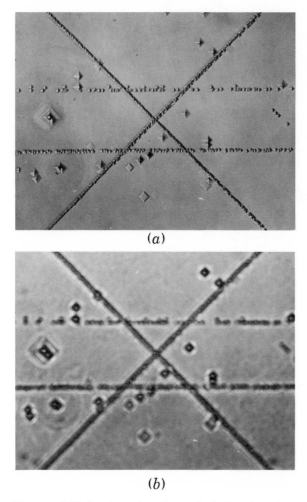

(a)

(b)

Fig. 6.4   Glide bands nucleated at an inclusion: (a) microscope focused on glide bands at surface; (b) microscope
focused on inclusion (black dot at center of cross) beneath
surface. Magnification = 25×.

description of the state of the material is compact enough to be useful.
Therefore, no attempt will be made here to keep track of the details.

WHEN ALL DISLOCATIONS ARE MOBILE (APPROACH TO STEADY STATE)

At high temperatures (above the Debye temperature), dislocation interactions do not permanently immobilize individual dislocations, because atomic

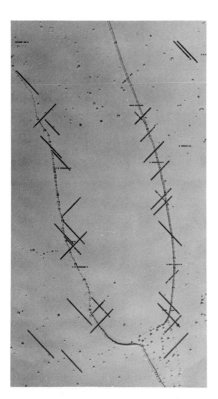

Fig. 6.5   Nucleation of glide bands at subgrain boundaries: crystal of neutron-irradiated LiF given a 1-sec 20 kg/mm² stress pulse.   Magnification = 100×. Specimen was selectively etched before and after the stress was applied.

diffusion occurs which allows static configurations to "dissolve." This leads to somewhat different rate behavior than that found when permanent obstacles form.

Let s_0 be the rate at which sources create dislocation flux in a specimen and $\alpha = mv$ be the breeding rate. These coefficients result in an ever-increasing dislocation density with time (or strain), but this means that the probability of an encounter between two or more dislocations increases also. The term *encounter* refers to many possibilities, including:

1. Parallel lines meet and annihilate one another.
2. Parallel lines meet and stalemate one another.

3. A line meets its "image" at a free surface and becomes annihilated.
4. Parallel lines meet and react to form an immobile line.
5. Nonparallel lines intersect and partially annihilate or stalemate one another.

Since the probability of a pair-wise encounter depends on the presence of two dislocations in a given region at a given time and the probability for each individual is proportional to N, the probability for a pair is proportional to N^2.

Eventually, since the rate of encounters increases more rapidly with N than does the rate of creation, the net rate of change of the dislocation density becomes zero, and the density reaches a steady-state value (Johnston and Gilman, 1959). This can be expressed analytically by letting β be the attrition coefficient (and neglecting encounters between more than two lines). Then the kinetic equation is

$$\dot{N} = s_0 + \alpha N - \beta N^2 \tag{6.9}$$

and this can be integrated to obtain N as a function of time (Webster, 1966). The result is

$$N = \frac{A - B\theta e^{-\Phi t}}{1 + \theta e^{-\Phi t}} \tag{6.10}$$

where $\Phi = (\alpha^2 + 4s_0\beta)^{\frac{1}{2}}$

$$A = \frac{1}{2}\frac{\alpha}{\beta}\left(\frac{\Phi}{\alpha} + 1\right)$$

$$B = \frac{1}{2}\frac{\alpha}{\beta}\left(\frac{\Phi}{\alpha} - 1\right)$$

N_0 = initial density

$$\theta = \frac{A - N_0}{B + N_0}$$

It may be seen that $N \to N_0$ for $t = 0$ and it approaches A as $t \to \infty$. Then, if α^2 is large compared to $4s_0\beta$, the large time limit of N is simply the ratio of the breeding and attrition coefficients α/β. On the other hand, if numerous sources are present but not much cross-glide breeding occurs, then the limiting value of N is $\frac{1}{2}(s_0/\beta)^{\frac{1}{2}}$.

BEHAVIOR WHEN SOME DISLOCATIONS BECOME IMMOBILIZED (STRAIN–HARDENING)

At low temperatures (below approximately the Debye temperature) some dislocations become permanently immobilized. The remaining ones continue to multiply and interact. Sometimes a moving one will interact with another moving one and sometimes with an immobilized one. The trend of the behavior may be deduced by inference from creep observations.

For a variety of materials, it is observed that creep at low temperatures begins rapidly and becomes progressively slower. It does not approach a constant rate, rather a constantly decreasing rate. Typically a "logarithmic creep law" is observed with the strain given by

$$\epsilon_s = c \ln t \tag{6.11}$$

which means that the creep rate is inversely proportional to the time:

$$\dot{\epsilon}_s = \frac{c}{t} \tag{6.12}$$

and, if (6.11) and (6.12) are combined,

$$\dot{\epsilon}_s \sim e^{-\epsilon_s/c} \tag{6.13}$$

In terms of dislocation motion, assuming that the mean velocity is constant at constant stress, Eq. (6.13) means that the mobile density must decay exponentially as the strain increases. Since the total population is known to remain constant or increase, it must be concluded that only a fraction of the total remains mobile. Let f be this fraction (which may be viewed equivalently as the fraction of time that a mean dislocation moves with the velocity it would have in the unstrained crystal). Then,

$$N_m = fN_t \tag{6.14}$$

The variation of the fraction f with strain may be found by considering how it must change. First, it is expected to decrease. Next, incremental decreases in f will be proportional incremental increases in N_t, since these lead to greater probabilities for encounters. Finally, only the mobile part of an incremental change in N_t will participate in further changing f. Thus

$$df \sim -f \, dN_t \tag{6.15}$$

But, from Eq. (6.8), $dN_t \sim d\epsilon$, so the equation for the change of f becomes

$$df = -\Phi f \, d\epsilon_p \tag{6.16}$$

where Φ is the attrition coefficient. Upon integration,

$$f = e^{-\Phi \epsilon_p} \tag{6.17}$$

and the mobile dislocation density is (compare Eq. 6.13)

$$N_m = (N_0 + M\epsilon_p)e^{-\Phi \epsilon_p} \tag{6.18}$$

Figure 6.6 illustrates the form of the behavior.

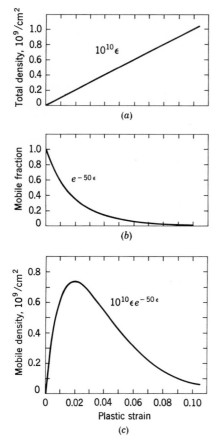

Fig. 6.6 Schematic drawings of the changes with increasing strain of: (a) total dislocation density; (b) mobile fraction; (c) mobile dislocation density.

For strains less than the maximum in Fig. 6.6c, the rapidly increasing dislocation density causes strain-softening, or *yielding*, in a manner that will be discussed in detail later. After the maximum has been exceeded, the decreasing mobile population results in strain-hardening, because a given strain rate can only be maintained by increasing the stress to raise the mean velocity.

Since most dislocation interactions become less inhibitory as the stress rises, the rate of attrition will decrease with increasing stress. For short-range interactions, the coefficient Φ will decrease inversely with the stress, so the dependence may be written as follows in terms of a *hardening coefficient H:*

$$\Phi = \frac{H}{\sigma_s} \tag{6.19}$$

This relation will vary among interaction mechanisms, but the given one describes typical situations.

Equation (6.18) results only if mobile dislocations are initially present in the material. This is usually true, but it can sometimes be avoided through careful crystal growth or annealing. For very small strains, a different description is then needed, but Eq. (6.18) with $N_0 = 0$ will become satisfactory as soon as a small amount of plastic strain has taken place.

It is interesting that there is algebraic equivalence between the viewpoint above and the one which takes the average dislocation velocity to be the parameter which changes with strain. This is easily shown by multiplying Eqs. (6.17) and (5.66), and incorporating (6.19) and the saturation velocity to obtain

$$\langle V \rangle = fv = v^* \exp \left[-(D + H\epsilon_p)/\sigma_s \right]$$

which causes the characteristic drag stress to become strain-dependent, instead of the mobile population.

6.3 VERY LARGE STRAIN BEHAVIOR

If Eq. (6.19) is substituted into (6.18) and the strain is large so that $N_0 \ll M\epsilon_p$, then (6.19) becomes the following after taking logarithms of both sides and solving for the stress:

$$\sigma_s = \frac{H\epsilon_p}{\ln (M\epsilon_p/N_m)} \tag{6.20}$$

Thus the stress is nearly proportional to the plastic strain. This cannot continue indefinitely, however, so a different description is needed when ϵ_p is very large, say more than 30 percent.

At large strains, crystals contain profuse quantities of dislocation dipoles (Gilman, 1964). This is observed in salt and metal crystals (Fourie and Murphy, 1962), as illustrated by Fig. 6.7. Also, it is expected, because dipoles do not have long-range stress fields so they are stable with respect to a pair of monopoles. Furthermore, they interact to form quadrupoles and other higher-order multipoles (Chen, Gilman, and Head, 1964). Because of these interactions, the dipole concentration eventually saturates. Kratochvil (1968) has discussed the effects of dipole clusters on stress-strain curves.

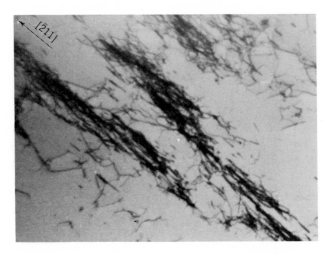

Fig. 6.7 Tangled dislocation dipoles in a strained copper crystal. Magnification = 30,000✕. (*Courtesy of J. T. Fourie.*)

For small strains the dipole concentration N_d is proportional to the strain, so the rate of change with strain $dN_d/d\epsilon$ is a constant, say μ_0. Then many dipoles begin to be formed by the interaction of monopoles with already existing dipoles, so $dN_d/d\epsilon$ becomes proportional to the instantaneous concentration. As the concentration rises further, dipole-dipole annihilation begins, so $dN_d/d\epsilon$ decreases in proportion to N_d^2. All this may be written

$$\frac{dN_d}{d\epsilon} = \mu_0 + \xi N_d - \chi N_d^2 \tag{6.21}$$

This has the same form as Eq. (6.9) for monopoles, and its solution is essentially the same, except that strain now plays the role of time.

The incremental stress associated with strain-hardening is a monotonic function of N_d, so saturation of the dipole concentration leads to a leveling off of the stress-strain curve at large strains.

REFERENCES

Alexander, H., and P. Haasen: *Solid State Phys.*, **22**: 28 (1968).

Brenner, S.: *J. Appl. Phys.*, **27**: 1484 (1956).

Chen, H. S., J. J. Gilman, and A. K. Head: *J. Appl. Phys.*, **35**: 2502 (1964).

Clarebrough, L. M., D. Michell, and A. P. Smith: *Phys. Stat. Solidi.*, **21**: 369 (1967).

Cottrell, A. H.: "Dislocations and Plastic Flow in Crystals," Oxford University Press, Fair Lawn, N.J., 1953.

Fourie, J. T., and R. J. Murphy: *Phil. Mag.*, **7**: 1617 (1962).

Gilbert, A., B. A. Wilcox, and G. T. Hahn: *Phil. Mag.*, **12**: 649 (1965).

Gilman, J. J.: *Discussions Faraday Soc.*, **38**: 123 (1964).

——: *J. Appl. Phys.*, **34**: 1584 (1959).

Johnston, W. G., and J. J. Gilman: *J. Appl. Phys.*, **31**: 632 (1960).

Ibid., **30**: 129 (1959).

Koehler, J. S.: *Phys. Rev.*, **86**: 52 (1952).

Kratochvil, P.: *Acta Met.*, **16**: 1023 (1968).

Li, J. C. M.: *J. Appl. Phys.*, **32**: 593 (1961).

Seitz, F., and T. A. Read: *J. Appl. Phys.*, **12**: 100 (1941).

Webster, G. A.: *Phil. Mag.*, **14**: 775 (1966).

7

MACROSCOPIC PLASTICITY

There are many possible descriptions of macroscopic plastic flow. An optimum one should be neither so simple that it covers only a few special cases, nor so complex that it is not useful for designing machines or gaining physical insight to plastic phenomena. The former condition requires that microscopic parameters be included in the description and also that time play a central role. Microscopic parameters must be included, because it is well-known that not only the external shape changes during plastic flow, but also the internal structure, especially the distribution of dislocations. Time must play a central role because plastic flow is a dissipative process. In the case of metals, about 95 percent of the plastic work gets converted into heat, and only about 5 percent is stored in the material in the form of additional dislocations and other defects.

The second condition of the paragraph above requires that the description be a statistical one, because the internal structure is exceedingly complex. The most simple type of statistical theory is one in which only the average values of the parameters are considered. More elaborate treatments deal with various distribution functions and their moments. The object here is to examine only the consequences of the most simple of treatments in which average parameters are used. It will be shown that this is adequate for most purposes.

From this point of view, no attempt is made to account for each dislocation that traverses a specimen. Instead, just as the kinetic theory of gases obtains statistical averages that can be related to macroscopic properties, the theory smears all the dislocations into a homogeneous distribution that is described by a strain-dependent density function and a stress-dependent average velocity.

The starting point is a strain-rate equation which is combined with various equilibrium and boundary conditions in order to describe the results of creep, stress strain, and plane elastoplastic waves in materials. The function of the strain-rate equation is to connect stress, strain, and time in a way that is determined by the behavior of dislocations, by their distributions, and by their interactions. It plays the role in plastic flow theory that Hooke's law plays in elastic deformation theory.

7.1 PROPERTIES OF MEAN LOOPS

A general form for the strain-rate equation has already been given as Eq. (5.13) in Sec. 5.1. It was also shown there that the rate equation is not highly sensitive to the microscopic distribution of mobile dislocations. This makes it quite reasonable to replace complicated sums of the parts of the distribution by average values. In other words, although the internal structure of a crystal contains a complex variety of dislocation-loop shapes and sizes during plastic flow, the behavior of an arbitrary loop does not differ much from that of a mean loop, so the character of a mean loop determines most of the behavior of the distribution.

It is usually more convenient to measure dislocation-line fluxes than loop concentrations, so the relations between these quantities need to be discussed. Suppose the concentration of mean elliptical loops is ξ (loops/cm³). Then, since the mean loop perimeter p is $2\pi \sqrt{kj}$ (see Sec. 5.1, Fig. 5.5), the total line length per unit volume is $2\pi\xi \sqrt{kj}$. If $N_\perp$ is the flux of lines through an arbitrary plane normal to $\mathbf{b}$ and perpendicular to the glide plane and $N_\parallel$ is the parallel flux, then to a first approximation,

$$\xi p = \sqrt{N_\perp N_\parallel} = \bar{N} \tag{7.1}$$

where $\bar{N}$ is the geometric mean of the two fluxes.

The conclusion of the above is that the dislocation structure of a crystal can be characterized by the geometric mean of two fluxes $N_\perp$ and $N_\parallel$. There are special cases (Cottrell, 1963), such as that of very small crystals or very low dislocation density where this is not a good approximation, but for most cases it will be good.

Another special case is that of very soft crystals in which the dislocation's stress fields may couple them so strongly that correlation effects become important and it is necessary to consider collective rather than individual behavior (Gilman, 1968).

After the mean mobile dislocation density in a crystal has been specified, the plastic strain rate depends only on the expansion rate of the mean loop.

If an incremental advance along the outward normal vector is $ds = \bar{v}\,dt$, where $\bar{v}$ is the geometric mean of the velocities of the edge (v_e) and screw (v_s) components,

$$\bar{v} = \sqrt{v_e v_s} \tag{7.2}$$

then the increment in area will be

$$d\alpha = p\,\sqrt{v_e v_s}\,dt = 2\pi\,\sqrt{kj}\,\sqrt{v_e v_s} \tag{7.3}$$

but $j = v_e t$, and $k = v_s t$, so

$$\dot{\alpha} = 2\pi v_e v_s t \tag{7.4}$$

which is the result obtained previously from a different viewpoint, and this shows that the geometric mean velocity is the appropriate parameter.

Dislocation movements are resisted by substantial drag forces in most crystals. Furthermore, since the effective masses of dislocations are only about 1 atomic mass per atom plane, they accelerate to the steady state in a time that is short compared with feasible loading times. Therefore, it is the mean steady-state velocity that is nearly always most important. In exceptional cases, it is necessary to begin with the strain-acceleration equation, and get the strain rate by integration.

7.2 AXIAL STRAIN RATE IN TERMS OF GLIDE

In the presence of a three-dimensional system of stresses, plastic strain rates can be expected to appear along three principal directions; some suggestions for describing this situation have been proposed (Mura, 1965; Bodner, 1968). A satisfactory general description is very difficult, however, because plastic flow leads to textures of dislocation lines, and because the results of dislocation interactions depend on the ratios of the principal strain rates. Therefore, no three-dimensional description will be attempted here. Instead the discussion will concentrate on one-dimensional situations; first, the axial extension or compression of a rod. Figure 7.1 illustrates the geometry.

The axial strain ϵ is defined as δ/L, where δ is the extension and L is the length. Thus the axial strain rate is

$$\dot{\epsilon} = \frac{1}{L}\frac{d\delta}{dt} \tag{7.5}$$

If A_g is the area of the glide plane and α is the area of the ith loop that lies on the plane, then the length change caused by expansion of the ith loop is

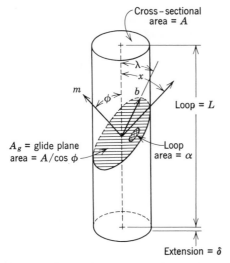

Fig. 7.1 Schematic geometry of a particular glide system in a cylindrical rod with stress applied parallel to its axis.

proportional to the fractional area swept out by motion of the loop and the component of the Burgers displacement in the axial direction:

$$d\delta_i = \left(\frac{d\alpha_i}{A_g}\right) b \cos \lambda \tag{7.6}$$

but the glide-plane area is related to the cross-sectional area by $A = A_g \cos \phi$, so the expression for the strain rate caused by the ith loop is

$$\dot{\epsilon}_i = \frac{\dot{\alpha}_i b \cos \lambda \cos \phi}{AL} \tag{7.7}$$

and the total rate is the sum over all the loop rates:

$$\dot{\epsilon} = \sum_1^M \left(\frac{b \cos \lambda \cos \phi}{V}\right) \dot{\alpha}_i$$
$$= \left(\frac{M}{V}\right) \bar{\dot{\alpha}} b \cos \lambda \cos \phi \tag{7.8}$$

and M/V is the loop density ξ, while $\bar{\dot{\alpha}}$ equals the perimeter of the mean loop times the average velocity $\bar{v}$ of the loop's line in the direction of the outer normal to the line. Therefore,

$$\dot{\epsilon} = \xi b p \bar{v} \cos \lambda \cos \phi \tag{7.9}$$

or, from Eq. (7.1) above,

$$\dot{\epsilon} = bN\bar{v} \cos \lambda \cos \theta \qquad (7.10)$$

In some cases, more than one glide system may be active, so the strains produced by the various systems must be added together to get the total. Thus, if the strain rate for this jth system is

$$\dot{\epsilon}_j = (bNpv)_j$$

the total is

$$\dot{\epsilon} = \sum_1^M \dot{\epsilon}_j \cos \lambda_j \cos \phi_j \qquad (7.11)$$

For large strains, λ_j and θ_j are functions of the strain, as shown in the section on crystal plasticity. But these angles may be taken to be constant for small strains or when a multiplicity of glide systems allows the most active system always to be close to the system with maximum shear stress, where $\lambda = \phi = 45°$ and $\cos \lambda \cos \phi = \frac{1}{2}$. Otherwise,

$$\cos \phi_j = \frac{\cos \phi_j^{\circ}}{1 + \epsilon} \qquad \sin \lambda_j = \frac{\sin \lambda_j^{\circ}}{1 + \epsilon} \qquad (7.12)$$

7.3 FLOW AT CONSTANT STRESS

This is the most simple dynamical situation because the average dislocation velocity may be taken to be constant (although equivalently it may be taken as variable with a constant fraction of the dislocation population being mobile). Then the only microscopic variable is the density of mobile dislocation lines. There are three homogeneous (and one inhomogeneous) cases to be considered. These are:

1. Steady-state creep
2. Transient approach to steady-state creep
3. Transient creep attenuated by strain-hardening
4. Propagation of inhomogeneous-plastic-flow bands (Luders bands)

They will be discussed briefly in turn.

STEADY-STATE CREEP

In order for steady-state creep to occur, the structure of the material must remain statistically constant as time passes. Therefore, the interactions that cause strain-hardening must dissolve at a rate that is adequate to prevent additional hardening. Since the hardening is caused by an increase in the total dislocation density, this means that this quantity remains constant; that is, the multiplication rate equals the attrition rate, as discussed in Sec. 6.2.

The multiplication rate is determined by the rate of cross glide, plus the rate at which fixed sources operate. The attrition rate is determined by the rate at which mass diffusion can dissolve edge-dislocation dipoles, plus the rate of surface egress. A detailed discussion of these processes is outside the present scope, but they have been reviewed in a general way recently by Garafalo (1965), and specific mechanisms have been discussed by Li (1966).

TRANSIENT APPROACH TO STEADY-STATE CREEP

Immediately after a stress has been applied to a material, a period of transient flow occurs while the steady-state structure is being established. Equations (6.9) and (6.10) of Sec. 6.2 can be used to describe this period. The glide strain rate is then

$$\dot{\epsilon} = bvN(t) \tag{7.13}$$

which can be integrated to obtain the dependence of the transient creep strain on time. This has been done for the case of copper by Akulov (1964), and for austenitic stainless steel by Li (1963). Figure 7.2 shows the results obtained by Li. Other results are described in the book by Garafalo (1965).

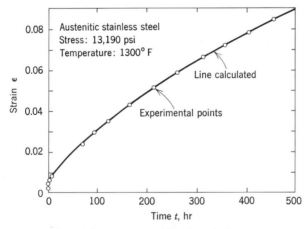

Fig. 7.2 Comparison of measured and analytic creep curves for an austenitic stainless steel. (*After Li*.)

TRANSIENT CREEP ATTENUATED BY STRAIN–HARDENING

Strain-hardening can be described in various ways and has been by various authors. If this is done carefully, the different methods are essentially equivalent, but the present author prefers the method described in Eqs. (6.18) and (6.19) of Sec. 6.2. This leads to the following rate equation (at constant stress):

$$\dot{\epsilon}_p = gbv(\sigma_s)[(N_0 + M\epsilon_p)e^{-H\epsilon_p/\sigma_s}] \tag{7.14}$$

where g is the geometric factor. This can be integrated to yield a relation between time and plastic strain (Gilman, 1965):

$$t = \left(\frac{Ke^{-KN_0}}{gvb\Phi}\right)[Ei(KN_0 + \Phi\epsilon_p) - Ei(KN_0)] \tag{7.15}$$

where $Ei(x)$ is the tabulated exponential integral function $= \int_{-\infty}^{x} \frac{e^x}{x}\, dx$; $\Phi = H/\sigma_s$; $K = \Phi/M$. This equation is compared with a measured curve for a lithium fluoride crystal in Fig. 7.3. All the constants except H were

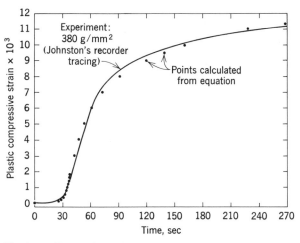

Fig. 7.3 Comparison of calculated and experimental creep curves for a LiF crystal.

determined approximately through independent measurements, so the close agreement here between theory and experiment is especially significant.

The transient-creep behavior depends on the initial dislocation density N_0, as shown in Fig. 7.4. The smaller the initial density, the greater the incubation time before rapid creep begins. For large values of N_0, as would

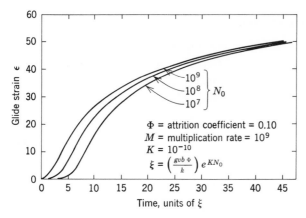

Fig. 7.4 Effect of initial dislocation density on computed transient-creep curves.

be expected for polycrystalline metals, the incubation period shrinks to zero.

A second important parameter is the strain-hardening, or attrition, coefficient, and Fig. 7.5 illustrates its effect. It determines both the maximum

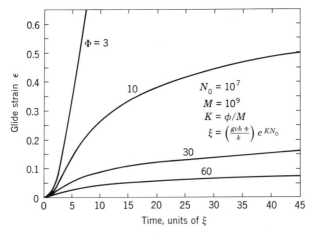

Fig. 7.5 Effect of strain-hardening coefficient on computed transient-creep curves.

creep rate and the total strain at a given time if all other factors remain constant.

Equation (7.15) has another important feature. The asymptotic limit of $Ei(x)$ for large values of x is

$$\ln Ei(x) \simeq x - \ln x \simeq x \qquad x \gg 1$$

therefore, for large strains, Eq. (7.15) becomes

$$\epsilon_p = \frac{1}{\Phi} \ln \frac{gvb\Phi t}{K} \tag{7.16}$$

which is the commonly observed logarithmic creep law (Garafalo, 1965).

Webster (1966) has postulated that, for small strains, the dependence of the mean dislocation velocity on dislocation density is linear and may be written

$$v = v_0 - kN \tag{7.17}$$

where v_0 is the initial velocity and k is a constant. Since N is proportional to ϵ, this leads to strain-hardening and is a reasonable approximation to a more general function $v(N)$. It has the advantage that it can be substituted into Eq. (7.13) together with Eq. (6.9) of Sec. 6.2 to obtain an explicit creep equation:

$$\epsilon = At + D \ln [B + (1 - B)e^{-\Phi t}]$$
$$+ c(1 - B)(1 - e^{-\Phi t})[B + (1 - B)e^{-\Phi t}]^{-1} \tag{7.18}$$

which can take many forms, depending on the values of the constants which all have definite physical meanings. The reader is referred to Webster's paper for the definitions of the constants.

Applications of Eq. (7.18) to data for two quite different materials (Al_2O_3 and a nickel-base superalloy) are shown in Fig. 7.6.

Another approach has been that of Haasen and coworkers (Alexander and Haasen, 1968) who have concentrated on the cases of germanium and silicon. They have treated the hardening process in terms of the development of a *back stress* that is caused by the elastic interactions of dislocations and therefore is proportional to the square root of the dislocation density. This back stress opposes the applied stress, so the mean dislocation velocity decreases as the density increases. It is assumed that all of the dislocations are mobile. It may be noted that this treatment is different in philosophy but statistically equivalent to the previous approaches. It allows an excellent correlation of quite different creep curves to be made, as shown in Figs. 7.7 and 7.8 (Reppich, Haasen, and Ilschner, 1964). Furthermore, the authors have shown that the constants needed to correlate the macroscopic creep tests are self-consistent with independent microscopic measurements. Thus the validity of Eq. (7.13) is well-verified.

The same approach has been used independently to correlate data for Ge with good results by Govorkov, Indenbem, Papkov, and Regel (1964).

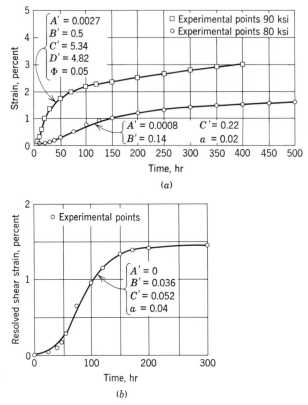

Fig. 7.6 Comparisons of experimental data with computed creep curves (*after Webster*): (*a*) nickel-base superalloy MAR M-200 at 760°C; (*b*) aluminum oxide at 1100°C and stress of 260 kg/cm². (*Data from Wachtman.*)

PROPAGATION OF INHOMOGENEOUS PLASTIC FLOW

If a slender rod or ribbon of plastic material is loaded in tension by some fixed amount just sufficient to initiate plastic flow, the flow does not begin everywhere along the rod. Instead it begins at some place where a stress concentration exists, such as the vicinity of an end grip. Locally the material extends plastically along the length (with a corresponding contraction in the transverse direction). Thus an interfacial zone is created between material that remains essentially elastic and a locally deformed region. The interfacial zone, or *Luders band front*, then tends to propagate along the length of the specimen until the entire rod has undergone plastic deformation.

Some of the most extensive measurements of this effect have been made

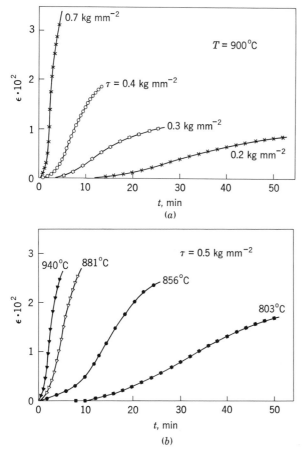

Fig. 7.7 Creep curves for silicon crystals as measured:
(a) various stresses at 900°C; (b) various temperatures for
stress of 0.5 kg/mm². (*After Reppich, Haasen, and Ilschner.*)

by Butler (1962) for steels in which the effect is particularly pronounced.
Figure 7.9 shows some of his data for the positions of some fronts as a function of time at constant stress. The data indicate that, indeed, propagation
occurs at a constant velocity which is a function of applied stress.

Hahn (1962) has pointed out that a creep law can be used to describe the
shape of a configuration that moves at constant velocity because continuity
must be preserved. This means that the following relation between strain
rates and strain gradients is valid:

$$\frac{\partial \epsilon}{\partial t} = V_B \frac{\partial \epsilon}{\partial l} \qquad (7.19)$$

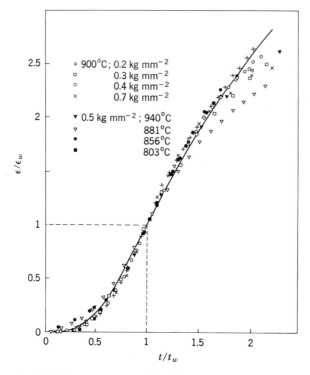

Fig. 7.8 The creep curves for silicon (Fig. 7.7) correlated to a single master curve. (*After Reppich, Haasen, and Ilschner.*)

Thus an expression for the strain rate can be converted into an expression for the strain gradient. The latter can be integrated to obtain the shape of a specimen. For large transverse strains, a correction must be introduced to account for the stress change that occurs when the front is crossed, but this is not necessary for small strains. It may be concluded that a band front will have the same qualitative shape (strain versus distance) as the curve in Fig. 7.3 with time replaced by position.

Since a Luders band propagates with a constant shape, the average dislocation configuration within it is fixed, even though the details of the configuration may not be known. This allows average dislocation velocities to be measured in a relative way, because the dislocation velocity at a particular cross section (where $\epsilon = \epsilon_0$) is given by (Gilman, 1965)

$$V_s = V_B Z \left(\frac{\partial \epsilon}{\partial l} \right)_{\epsilon_0} \tag{7.20}$$

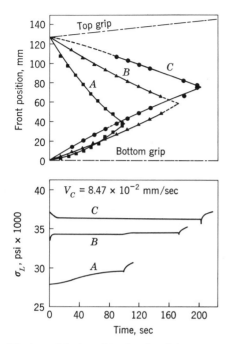

Fig. 7.9 Motion of Luders band fronts in steel wires (0.025 wt% C) at nearly constant stress. Grain sizes: A—24μ; B—15μ; C—13μ. Crosshead speed = V_c. (*After Butler.*)

where Z is a constant. The last factor in this equation is important because the shape of the band front depends on the velocity, as pointed out by Hart (1955). If the band velocity is known as a function of stress together with the shape factor, then the variation of the dislocation velocity v_s with stress becomes determined through Eq. (7.20). Furthermore, since the shape factor $(\partial E/\partial l)_{E_0}$ changes slowly with the band-front velocity, whereas the velocity changes rapidly with the applied stress, it is usually a good approximation to take v_s proportional to V_B. Then Butler's measurements of $V_B(\sigma_s)$, shown in Fig. 7.10, indicated how V_s depends on stress, and carbon content, and grain size. Note that increased carbon content increases the effective viscosity, as does decreased grain size.

Butler further found that the stress for band-front propagation at a given velocity is proportional to the reciprocal square root of the grain size (grain diameter = d). Thus the drag stress has the following grain-size dependence: $D = D_0 + D_1/\sqrt{a}$, where D_0 is the value for a monocrystal.

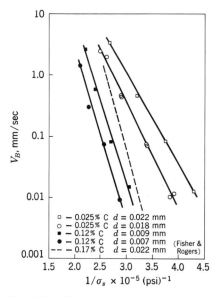

Fig. 7.10 Showing that log V_B is proportional to $-D/\sigma_s$ for steel wires of various compositions and grain sizes. (*After Butler.*)

This method has been applied also to nylon fibers by Dey (1967), and some of his results are shown in Fig. 7.11, where some band-front velocities are shown to depend on stress in qualitatively the same way they do for the steels shown in Fig. 7.10. This similarity is encouraging to the idea that plastic flow in all materials can be described in a unified way in terms of suitable strain-rate equations.

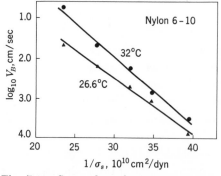

Fig. 7.11 Stress dependence of deformation-band velocities in nylon 6-10 at two temperatures. (*After Dey.*)

The dynamics of the development of the unstable flow which causes Luders bands to form has been discussed by Campbell (1968).

7.4 THE TENSION TEST (CONSTANT DISPLACEMENT RATE)

Of all mechanical tests for structural materials, the tension test is the most common. This is true primarily because it is a relatively rapid test and requires simple apparatus. It is not as simple to interpret the data it gives, however, as might appear at first sight. Tensile testing is done with various types of machines. The easiest to analyze in detail is the constant-displacement-rate type in which a rigid crosshead moves at a fixed speed and one end of a specimen is attached to it and pulled. The other end of the specimen is attached to a stiff spring whose deflection is measured to determine the axial force that acts on the specimen at any instant. The force is recorded as a function of the crosshead displacement (or the strain in the specimen if a separate strain gage is used).

For many years the stress-strain curves that were obtained from such tests were interpreted in an essentially static fashion; for example, the theory of the yield point of steel (Cottrell, 1948). In this theory solute atoms lock the dislocations in place until a critical stress (yield stress) is reached. Then the dislocations break free and produce flow.

Static theories are partially successful because tension test results are not very rate-sensitive for many structural materials, especially in contrast to the effects of heat treatment, cold working, or alloying. At high rates the sensitivity increases, however, and cannot be ignored.

If a machine could impose a constant strain rate on a specimen, it would be possible to simply record the stress as a function of time to obtain a stress-strain curve. This is not the case, however, because the total strain rate equals the sum of the elastic and plastic strain rates:

$$\epsilon_T = \epsilon_P + \epsilon_E \tag{7.21}$$

Therefore, a constancy of ϵ_T does not necessarily mean that ϵ_P is constant; only the sum of ϵ_P and ϵ_E is constant.

A constant-displacement-rate machine applies a constant total strain rate that is the sum of:

1. Elastic strain rate in specimen
2. Plastic strain rate in specimen
3. Strain rate resulting from the elastic compliance of the machine, especially the load-measuring spring

Therefore, at any instant of time there is some distribution of the total rate among these components. This distribution changes with time as the machine and the specimen interact.

In order to determine the distribution of strain rates, it is necessary to have an equation that describes the machine-specimen interaction (Gilman, 1956). A schematic diagram for a machine is shown in Fig. 7.12. An equation

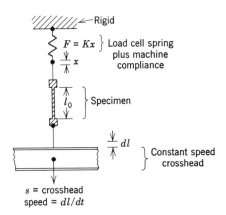

Fig. 7.12   Diagram of equivalent tensile testing machine.

for describing its behavior is found by equating the various displacements of the system at a particular time t. If the crosshead speed is S, then the total displacement is St. The force F on the specimen causes elastic displacements in the load-cell spring and the frame of the machine. If the total compliance of these components is K, then the machine displacement is F/K.

Suppose that the gage length of the specimen is l_0 and its cross-sectional area is A. Then the axial stress in it is $F/A = \sigma$, and if its Young's modulus is E, the stress induces a strain σ/E, so its elastic displacement is $\sigma l_0/E_0$.

The plastic displacement of the specimen equals the plastic strain ϵ_P in it times its length.

Equating the total displacement at time t to its parts yields

$$St = \frac{F}{K} + \frac{\sigma l_0}{E} + \epsilon_P l_0 \tag{7.22}$$

solving for the stress,

$$\sigma = \frac{St - \epsilon_P l_0}{A_0/K + l_0/E} \tag{7.23}$$

In this equation the denominator on the right is the total compliance constant of the machine and specimen. The numerator states that the stress increases as the crosshead speed does and relaxes as the integral of the strain rate increases.

To find the stress-strain curve, Eq. (7.23) must be solved simultaneously with a strain-rate equation such as Eq. (7.14). In very simple cases, this may be done analytically, but usually numerical methods are required. As an example, to illustrate the effect of the machine-specimen interaction, the case of prismatic glide in zinc crystals will be analyzed (Gilman, 1956). The flow law has a particularly simple form in this case because no strain-hardening occurs. At constant temperature,

$$\dot{\epsilon}_P = B\sigma^3 \tag{7.24}$$

but $\sigma = Cx$, with $C = K/A$. Also, the specimen compliance can be neglected, so the rate equation for the equivalent machine is

$$\frac{\dot{F}}{Kl_0} + C^3x^3 - \frac{s}{l_0} = 0 = \dot{x} + Dx^3 - S \tag{7.25}$$

with $\dot{F} = K\dot{x}$ and $D = C^3l_0$. Equation (7.25) may be transformed into a standard integral (letting $a^3 = S/D$):

$$Dt = \int_0^x \frac{dx}{a^3 - x^3} \tag{7.26}$$

and the problem is explicitly solved. The analytic result is compared with experimental points in Fig. 7.13.

Numerical solutions of the machine and material equations were obtained by Johnston and Gilman (1959), and for LiF crystals by Johnston (1962). They found that unstable yielding, delayed flow at constant stress, the general shapes of stress-strain curves, and the quantitative shapes at small strains could all be accounted for in this conceptually simple and unified way. Furthermore, a variety of material types can be described without changing the forms of the equations, although the details have varied among authors. Steels and other body-centered-cubic metals can be described, and their yield points may be predicted as shown by Hahn (1962). Also, delayed yielding and strain-aging phenomena in steel have been discussed by Hahn, Reid, and Gilbert (1962). Germanium and other covalent crystals behave quite differently at the microscopic level from metals and salts, yet Patel and Chaudhuri (1963) found that the macroscopic behavior of Ge can be successfully related to dislocation properties.

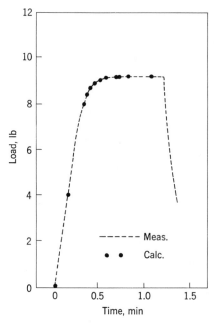

Fig. 7.13 Comparison of measured load-elongation curve with points calculated from Eq. (7.26) for a zinc crystal with tension applied parallel to its basal planes.

In order to represent the dependence of dislocation velocity on stress, several authors have used a power function

$$v \sim \sigma^m \tag{7.27}$$

where m varies between unity and as much as 200. This is satisfactory for moderate ranges of stress, but if the stress gets too large, the velocity exceeds sonic velocities, and this is not acceptable. Therefore Gillis and Gilman (1965) performed calculations in which the velocity-stress function was

$$v(\sigma_s) = v^* e^{-D/\sigma_s} \tag{7.28}$$

where v^* is the terminal or limiting velocity and D is the drag stress. This function has the interesting and useful property that it sets a finite yield stress below which the velocity is relatively small and above which it is relatively large. This property is illustrated by Fig. 7.14 which shows the reduced velocity v/c (here $v^* = c =$ sonic velocity) as a function of the reduced stress $x = \sigma_s/D$. The critical stress for fast motion is defined by the inflection point, where $\partial^2 v / \partial x^2 = 0$.

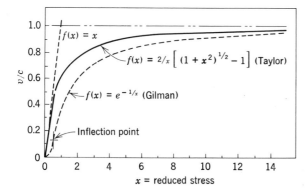

Fig. 7.14 Graphs of functions used to describe the dependence of steady-state dislocation velocities on stress.

Another property of this function is that it gives essentially zero velocity at zero stress. This may be seen clearly by studying the effective viscosity coefficient which determines the power loss during steady-state motion. It is defined as

$$\eta^* = b\frac{\sigma}{v} = \frac{b}{v^*}\,\sigma_s e^{D/\sigma}.$$ (7.29)

so the effective viscosity is infinite at zero stress.

Equation (7.28) accurately describes measured data for many structural materials that have rather definite yield points. However, high-purity metals and salts behave differently, because in them velocity is proportional to the stress at low stresses; that is, they are "newtonian." At high stress levels the velocity cannot continue to increase in proportion to the stress, because it will soon exceed the velocity of sound, as indicated in Fig. 7.14. Taylor (1968) has proposed that the reason it levels off is that the viscosity of the phonon gas in which a dislocation always moves increases at high velocities because of relativistic effects. By analogy with the behavior of an electron moving through a photon gas, he proposes that the damping coefficient varies as

$$B = \frac{B_0}{1 - v^2/c^2}$$ (7.30)

Then, since $\sigma_s b = Bv$, the dependence of the velocity on stress is

$$v = \frac{c}{x}[(1 + x^2)^{\frac{1}{2}} - 1]$$ (7.31)

where x is the reduced stress:

$$x = \sigma_s \left(\frac{B_0 c}{2b}\right)^{-1}$$

The *viscosity function* (7.31) gives a velocity proportional to the stress at low stresses in contrast to the *yield function* (7.28) which only gives an appreciable velocity when the stress approaches and exceeds the value $D/2$.

Structural materials usually have rather definite yield stresses, so they are best described by means of the yield function. Gillis and Gilman (1965) calculated a set of typical stress-strain curves as shown in Fig. 7.15. The

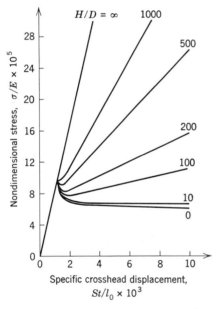

Fig. 7.15 Set of stress-strain curves given by numerically solving the tensile testing machine equation and the plastic strain-rate equation simultaneously. The yield function was used for $V(\sigma_s)$. The material constants were $b = 2.9 \times 10^{-8}$ cm; $\cos^2 \lambda_0 \cos^2 \Phi_0 = \frac{1}{2}$; $N = 10^2/\text{cm}^3$; $M = 10^9/\text{cm}^2$; $v^* = 2.8 \times 10^3$ cm/sec.

coordinates are nondimensional, as is the strain-hardening parameter. If the latter is infinite, the dislocations in the material never move, so the stress-strain curve is entirely elastic. At the other extreme of no strain-hardening, yielding occurs when the stress is large enough together with N

to make $\dot{\epsilon}_p l_0 \simeq S$, so that $\partial\sigma/\partial t = 0$. But N continues to increase in proportion to ϵ_p, so $\dot{\epsilon}_p l_0$ becomes greater than S and plastic relaxation occurs and the stress asymptotically approaches zero. If some strain-hardening occurs, however, the flow rate slows down as N becomes large, and the stress rises in approximate proportion to the plastic strain.

Depending on the values of the material and machine constants, it may be seen that the theory generates any of the known types of stress-strain curve.

Of particular interest is the dependence of the upper yield stress σ_y on the material and machine parameters. Figure 7.16 contains the results of some

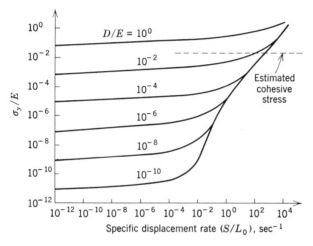

Fig. 7.16   Calculated dependence of upper yield stress on displacement rate for various values of the drag stress. (*After Gillis.*)

calculations by Gillis (1965) which show how the upper yield stress depends on the specific displacement rate S/l_0 (nominal strain rate). At small rates the upper yield stress is nearly independent of the rate, and this is consistent with experimental measurements. However, at higher rates the yield stress increases rapidly as the rate increases. This happens because the power loss caused by viscous forces increases with velocity.

The parameter that has the largest effect on the upper yield stress is the drag stress D because it appears in the argument of the exponential function whereas the other factors appear only as multipliers of the exponential. Indeed, the upper yield stress is proportional to the drag stress, whereas it changes only in proportion to the logarithms of the other parameters. The following factors tend to *decrease* the upper yield stress:

1. Increasing initial dislocation density
2. Decreasing machine stiffness
3. Increasing multiplication coefficient

Patel and Chaudhuri (1963) have given very clear evidence of the first of these effects, and some of their stress-strain curves are shown in Fig. 7.17.

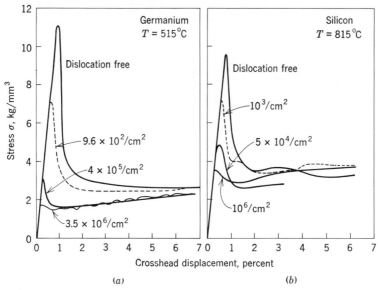

Fig. 7.17   Influence of initial dislocation density on the stress-strain curves of Ge and Si. Crosshead velocity $= 8.3 \times 10^{-5}$ cm/sec. (*After Patel and Chaudhuri.*)

7.5 PLANE-WAVE IMPACT

Impact phenomena provide a critical testing ground for theories of mechanical behavior. One reason is that experimental impact studies are difficult and expensive to perform, so a good theory that minimizes the number of required measurements becomes especially valuable. Another is that because of the short time scale involved and the destructive effects of strong impacts, it is feasible to measure only a few parameters during an impact experiment. A third is that stress, strain, and time all play a part during impact, so the full machinery of a theory is required to interpret the complex behavior that is observed. This imposes a critical test of the adequacy of the theory.

Until relatively recently, neither experimental techniques nor theoretical concepts had advanced sufficiently to demonstrate what happens when a piece of steel or other material is impacted. Available experimental techniques did not have sufficient simultaneous time and space resolution to determine some important features of the behavior. Also, the models that were used to describe the behavior were overly simplified. In part this was because more complex models were impractical before the advent of electronic computers.

The discussion here will concentrate on the plane-wave impact test because this is a quasi-one-dimensional and therefore more amenable to analysis than other tests such as the impact of slender bars. In a plane-wave impact a stationary specimen plate is struck by another plate that approaches the specimen at some moderate-to-high velocity v_0. At the instant of impact the moving plate imparts momentum to the specimen, and hence a propagating elastic wave. As the elastic wave propagates, the shear stresses in it can be relaxed by plastic flow (and most of the plastic work gets converted into heat). When the wave reaches the back surface of the specimen plate, this surface begins to move, and the velocity of this motion as a function of time indicates the shape of the propagating wave profile. Furthermore, if this is done for several plate thicknesses (so the wave travels further prior to the time at which its profile is measured), then its rate of attenuation can be obtained.

The first step in developing an understanding of impact is to consider the elastic waves that are present and how these result from the equations of motion.

ELASTIC WAVES

When a plane wave reaches a small volume element inside a plate, the particles in the element become displaced, as sketched schematically in Fig. 7.18. Since the element does not change in its transverse dimensions (by definition of a plane wave), both its volume and its shape change as the wave passes it. At position (2) is an element in its initial state ahead of the wave which has reached position (1) and caused displacements there. Note that the volume of the element at (1) has changed and that it has been sheared. The distortions cause stresses, and since the stress on one side of the element is not necessarily the same as on the other, a net force acts on the element that is proportional to the stress gradients. This force causes an acceleration of the element in accordance with Newton's laws of motion. Since there are three possible acceleration directions and nine stress gradients.

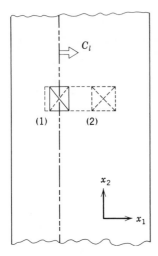

Fig. 7.18 Schematic effect of an elastic plane wave on a small volume element in a plate. At position (2) is an element that the wave has not yet reached. At (1) the particle displacements in the wave have changed the element.

the equations of motion for the element are (Kolsky, 1953):

ρ = mass density

μ_i = displacements

$$\rho \frac{\partial^2 \mu_1}{\partial t^2} = \frac{\partial \sigma_{11}}{\partial x_1} + \frac{\partial \sigma_{12}}{\partial x_2} + \frac{\partial \sigma_{13}}{\partial x_3}$$

x_i = coordinates

$$\rho \frac{\partial^2 \mu_2}{\partial t^2} = \frac{\partial \sigma_{21}}{\partial x_1} + \frac{\partial \sigma_{22}}{\partial x_2} + \frac{\partial \sigma_{23}}{\partial x_3} \qquad (7.32)$$

σ_{ij} = stress

t = time

$$\rho \frac{\partial^2 \mu_3}{\partial t^2} = \frac{\partial \sigma_{31}}{\partial x_1} + \frac{\partial \sigma_{32}}{\partial x_2} + \frac{\partial \sigma_{33}}{\partial x_3}$$

To relate these general equations to a particular material, it is necessary to use Hooke's law, and this takes the following form in the isotropic case:

$$
\begin{aligned}
\sigma_{11} &= \lambda\Delta + 2G\epsilon_{11} & \sigma_{12} &= G\epsilon_{12} = \sigma_{21} \\
\sigma_{22} &= \lambda\Delta + 2G\epsilon_{22} & \sigma_{13} &= G\epsilon_{13} = \sigma_{31} \\
\sigma_{33} &= \lambda\Delta + 2G\epsilon_{33} & \sigma_{23} &= G\epsilon_{23} = \sigma_{32}
\end{aligned}
\qquad (7.33)
$$

the volume change, or dilatation, is designated as Δ in these relations. Thus,

$$\Delta = \frac{\Delta V}{V} = \epsilon_{11} + \epsilon_{22} + \epsilon_{33}$$

$$= \frac{P}{K} \tag{7.34}$$

where K is the bulk modulus and P is the hydrostatic pressure. The elastic constants in Eqs. (7.33) are $\lambda =$ Lamé's constant and $G =$ shear modulus. It may be noted that $\lambda = K - (\frac{2}{3})G$, so the Lamé constant describes the effects of both volume and shape changes. This is why it is useful describing plane waves in which both types of distortion occur.

The stress-strain relations [Eq. (7.33)] may be used to transform the equations of motion [Eq. (7.32)] into equations that contain only derivatives of the displacements if the following definitions are used:

$$\epsilon_{ii} = \frac{\partial \mu_i}{\partial x_i}$$

$$\epsilon_{jj} = \frac{\partial \mu_j}{\partial x_i} + \frac{\partial \mu_i}{\partial x_j}$$

$$\nabla^2 = \sum_1^3 \frac{\partial^2}{\partial x_i^2}$$

so the equations of motion become

$$\rho \frac{\partial^2 \mu_1}{\partial t^2} = (\lambda + G) \frac{\partial \Delta}{\partial x_1} + G\nabla^2 \mu_1$$

$$\rho \frac{\partial^2 \mu_3}{\partial t^2} = (\lambda + G) \frac{\partial \Delta}{\partial x_2} + G\nabla^2 \mu_2 \tag{7.35}$$

$$\rho \frac{\partial^2 \mu_3}{\partial t^2} = (\lambda + G) \frac{\partial \Delta}{\partial x_3} + G\nabla^2 \mu_3$$

These equations govern the behavior of an elastic plane wave propagating in the x_1 direction as in Fig. 7.18. If the phase velocity of the wave is c, then any function of the following form is a solution of Eq. (7.35):

$$\mu_i = f(x_1 - ct) = f(\phi) \tag{7.36}$$

This can be verified by substitution. From this form of solution the following time and space derivatives are found:

$$\frac{\partial^2 \mu_i}{\partial t^2} = \frac{\partial \mu_i}{\partial \phi^2} \qquad i = 1, 2, \text{ or } 3$$

$$\frac{\partial^2 \mu_i}{\partial l_1^2} = \frac{\partial \mu_i}{\partial \phi^2} \qquad \frac{\partial^2 \mu_i}{\partial x_2^2} = 0 = \frac{\partial^2 \mu_i}{\partial x_3^2}$$

$$= \mu_i''$$

and the following equations are found when these relations are substituted into Eqs. (7.35):

$$\begin{aligned}
\rho c^2 \mu_1'' &= (\lambda + 2G)\mu_1'' \qquad &(a) \\
\rho c^2 \mu_2'' &= G\mu_2'' \qquad &(b) \\
\rho c^2 \mu_3'' &= G\mu_3'' \qquad &(c)
\end{aligned} \qquad (7.37)$$

Thus Eq. (7.36) is a solution, but there are two different propagation velocities, one given by Eq. (7.37a) and the other by Eqs. (7.37b) and (7.37c). A physical wave can have but one propagation velocity, so this means that two distinct types exist. The type described by Eq. (7.37a) has particle displacements in the propagation direction x (with $\mu_2'' = \mu_3'' = 0$), so it is a longitudinal wave with velocity:

$$c_l = \left(\frac{\lambda + 2G}{\rho}\right)^{\frac{1}{2}} \qquad (7.38)$$

The other type is transverse ($\mu_1'' = 0$) with velocity:

$$c_t = \left(\frac{G}{\rho}\right)^{\frac{1}{2}} \qquad (7.39)$$

If the wave is plane and in a plate that extends to large distances in the x_2 and x_3 directions, there can be no transverse displacements, so μ_2'' and μ_3'' equal zero, and only the longitudinal wave remains.

The expression for the longitudinal-wave velocity may be rewritten in terms of the bulk modulus K, which measures resistance to volume changes, and G, which measures shear resistance. Then,

$$c_l = \left(\frac{K + 4/3G}{\rho}\right)^{\frac{1}{2}} \qquad (7.40)$$

and it may be seen that a longitudinal wave moves faster than a pure pressure wave. Also, if a material has negligible shear resistance because it flows plastically at low stresses, then the effective value of G vanishes, and a pure pressure wave results.

STRESS INTENSITY INDUCED BY IMPACT

A specific impact situation is sketched schematically in Fig. 7.19. The initial events are independent of the thickness ratio of the two plates, so the chosen geometry is a convenient but general one for short times after impact.

At a time δt after impact, the elastic disturbance caused by the impact moves into the specimen a distance $\delta x = C_l \, \delta t$, as shown in Fig. 7.19$b$. This

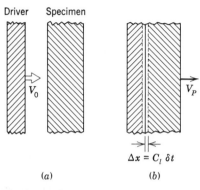

Driver   Specimen

V_0

V_P

$\Delta x = C_l \, \delta t$

(a)   (b)

Fig. 7.19 Schematic drawing of moving driver plate striking a specimen plate: (a) just before impact; (b) just after impact.

disturbance sets the particles in the region δx into motion. At the same time the particles in the driver plate must slow down in order to conserve momentum.

The displacement that is caused at the struck surface of the specimen is

$$\delta \mu_1 = \epsilon_{11} \, \delta x = \epsilon_{11} C_l \, \delta t$$

where ϵ_{11} is the elastic strain. Let the particle velocity in the specimen be

$$v_p = \frac{\delta \mu}{\delta t} = C_l \epsilon_{11}$$

then the strain in the specimen is

$$\epsilon_{11} = \frac{v_p}{C_l} \qquad \epsilon_{22} = \epsilon_{33} = 0 \tag{7.41}$$

If the driver and specimen are made of the same material, then momentum conservation requires that the velocity induced in the δx layer of the specimen be equal to $\frac{1}{2}$ driver velocity. Thus the v in Eq. (7.41) is $\frac{1}{2}$ the velocity of the driver, and the strain induced in the specimen is proportional to the impact velocity.

To find the stress intensity, Eqs. (7.33) may be used:

$$\sigma_{11} = (\lambda + 2G)\epsilon_{11}$$
$$\sigma_{22} = \lambda\epsilon_{11}$$
$$\sigma_{33} = \lambda\epsilon_{11}$$

(7.42)

and then the pressure in the wave is

$$P = \frac{1}{3}(\sigma_{11} + \sigma_{22} + \sigma_{33}) = \left(\lambda + \frac{2G}{3}\right)\epsilon_{11}$$

$$= K\frac{v_0}{2C_l}$$

(7.43)

while the maximum shear stress is

$$(\sigma_s)_{max} = \frac{1}{2}(\sigma_{11} - \sigma_{33}) = G\frac{v_0}{2C_l}$$

(7.44)

Both increase in proportion to the driver velocity.

When the shear stress is large enough rapid plastic flow occurs which reduces the effective shear stiffness G_{eff} to zero, and the wave tends to become a pure pressure disturbance. If, as a rough approximation, a fixed yield stress σ_y is defined above which arbitrarily rapid flow occurs, then the shear stiffness is negligible for impact velocities above

$$v_0^{crit} > \left(\frac{2\sigma_y}{G}\right)C_l$$

(7.45)

For higher velocities, the specimen tends to behave like a liquid; and since the maximum possible value of σ_y/G is about $\frac{1}{15}$, all solids tend to behave like liquids for impact velocities greater than $\sim C_l/10$ or about 5×10^4 cm/sec ($\sim$2000 ft/sec).

The elastic strains produced by very high-impact velocities become so large that Hooke's law [Eq. (7.33)] becomes invalid, and the stress-strain relations become nonlinear. For compressional waves this leads to the development of discontinuous displacements at the fronts of propagating waves, that is, to shock waves. The behavior of materials in this regime has been studied extensively (Rice, McQueen, and Walsh, 1958), but it is outside the scope of the present discussion which is restricted to impact velocities that cause a linear-elastic response.

PLASTIC RELAXATION OF AN ELASTIC WAVE

The microdynamical theory of plastic relaxation during impact was first developed by Taylor (1965), and his arguments are closely followed here. The total response (elastic plus plastic) of a small element that lies in the path of a plane wave is to be calculated. The material is polycrystalline and therefore taken to be elastically isotropic and plastically homogeneous. There is rotational symmetry about the x_1 axis, so

$$\sigma_{22} = \sigma_{33} \qquad \epsilon_{22} = \epsilon_{33} \tag{7.46}$$

The stresses and strains are taken to be positive in compression, so Hooke's law becomes

$$\sigma_{11} = (\lambda + 2G)\epsilon_{11}^e + 2\lambda\epsilon_{33}^e$$
$$\sigma_{33} = \lambda\epsilon_{11}^e + 2(\lambda + G)\epsilon_{33}^e \tag{7.47}$$

where the superscripts designate elastic strains, as distinguished from plastic ones ϵ_{ij}^p. Since the stresses arise only from the elastic strains, they have no superscripts.

The total strains are defined by

$$\epsilon_{11}^T = \epsilon_{11}^e + \epsilon_{11}^p = -\Delta$$
$$\epsilon_{33}^T = \epsilon_{33}^e + \epsilon_{33}^p = 0 \tag{7.48}$$

and since the plastic strains cause shape changes but no volume changes (in the first approximation), another condition exists,

$$\epsilon_{11}^p + 2\epsilon_{33}^p = 0 \tag{7.49}$$

It is important to realize here that plastic flow cannot change the primary displacement field of a simple plane wave because the primary displacement causes a volume change. Although plastic flow cannot change the volume of the distorted material, it can relax the shear strains within it and thereby reduce the shear-strain energy.

In a random polycrystal, the driving force for plastic flow is proportional to the maximum shear stress:

$$\sigma_s = \frac{1}{2} (\sigma_{11} - \sigma_{33}) \tag{7.50}$$

and the corresponding plastic shear strain is related to the principal plastic strains by

$$\epsilon^p = \frac{1}{2} (\epsilon_{11}^p - \epsilon_{33}^p) \tag{7.51}$$

From Eqs. (7.47–7.51) the following can be derived:

$$\sigma_{11} = (\lambda + 2G)\epsilon_{11}^T - \frac{8}{3} G\epsilon^p \tag{7.52}$$

$$\sigma_s = G(\epsilon_{11}^T - 2\epsilon^p) \tag{7.53}$$

To obtain a rate equation, partial time differentiation is applied to Eq. (7.52):

$$\dot{\sigma}_{11} = (\lambda + 2G)\dot{\epsilon}_{11}^{T} - \frac{8}{3}G\dot{\epsilon}^{p} \tag{7.54}$$

It is also instructive to examine the expression for the rate of change of the shear stress by substituting Eqs. (7.48) and (7.49) into (7.53) and then time differentiating. The stress rate is

$$\dot{\sigma}_{s} = G\left(\dot{\epsilon}_{11}^{e} - \frac{1}{2}\dot{\epsilon}_{11}^{p}\right) \tag{7.55}$$

which indicates that the stress rises as the elastic strain does, but is relaxed by the plastic strain rate. Equations (7.54) and (7.55) indicate the relaxing role played by the plastic strain rate, which is given in the following form by dislocation theory:

$$\dot{\epsilon}^{p} = bN_{m}(\epsilon^{p},\sigma_{s})\bar{v}(\epsilon^{p},\sigma_{s}) \tag{7.56}$$

where the mobile density N_{m} and the average velocity $\bar{v}$ are functions of the stress and plastic strain. Specific forms of these functions have been discussed previously.†

In order to take the inertia of the material into account, an equation of motion is needed. This may be written

$$\rho\frac{\partial v_{1}}{\partial t} + \frac{\partial \sigma_{11}}{\partial x_{1}} = 0 \tag{7.57}$$

with the first term being the inertial force, and the second the stress force. The mass density is ρ, and v_{1} is the mass particle velocity.

To preserve continuity of the material, the following additional condition is imposed:

$$\frac{\partial v_{1}}{\partial x_{1}} + \frac{\partial \epsilon_{11}}{\partial t} = 0 \tag{7.58}$$

The set of equations formed by (7.53), (7.56), (7.57), and (7.58) can be used to compute wave profiles that can be compared with experimentally observed ones. This has been done for small strains by Taylor (1965) and for

---

† Under impact conditions profuse twinning may occur in a material. If this occurs, Eq. (7.56) must be reinterpreted. One form that it might take is

$$\dot{\epsilon}_{\text{twinning}} = sp\dot{A}$$

where s = twinning shear

p = number of twins intersecting an arbitrary plane that passes through the specimen (No./cm²)

A = average rate of change of cross-sectional area per twin (cm²/sec) large strains by Wilkins. Special procedures are required because the elastic-wave equation (7.57) tends to give oscillating numerical results, especially when it is combined with Eq. (7.56), which usually has a nonlinear form. Successful computational methods have been described by Wilkins (1964).

Before the results of numerical computations are described, some experimental observations of wave profiles will be discussed so they may be compared with theoretical predictions.

TECHNIQUES FOR MEASURING WAVE PROFILES

The plane part of a propagating mechanical disturbance is necessarily immersed inside the material and therefore is quite inaccessible except to techniques such as flash x-radiography (Venables, 1964). Furthermore, the wave front may be very sharp with substantial changes occurring in about 10^{-8} sec (that is, a spatial spread of less than 5×10^{-3} cm). These specifications can only be satisfied by very fast sensing equipment applied at an external surface. It has become standard practice to measure the back-surface velocity, v_β (indicated in Fig. 7.19) as a function of time. Then Eq. (7.41) is used (with $v_\beta = 2v_p$) to determine the stress as a function of time. If the space distribution of the stress is desired, it can be obtained by multiplying the time intervals by the propagation velocity.

Early measurements of back-surface velocities used a series of contact pins that had slightly different lengths, so the wave reached their ends (and thereby closed the contacts) at slightly different times (Minshall, 1955). This method measures average velocities accurately but has insufficient time resolution for acceleration measurements. Later an optical wedge method was developed (Marsh and McQueen, 1960), but the first methods with adequate resolving power used capacitance changes. One of these (developed by Hughes, Gourley, and Gourley, 1962), used radio-frequency excitation of the capacitor, while the other (developed by Rice, 1961), used a static excitation voltage to maximize the time resolution. This latter method is outlined schematically in Fig. 7.20, which shows a polarized capacitor with the specimen as one electrode and a probe as the other. The output drives the grid of a cathode follower whose output is displayed on a fast oscilloscope and photographed. This method has also been used by Ivanov, Novikov, and Sinitsyn (1963).

Quartz transducers can also be used to measure surface velocities (Jones, Neilson, and Benedick, 1962), because the short-circuit current from a thin-quartz X-cut disk is directly proportional to the specimen-quartz interface stress for times less than the wave transit time in the disk.

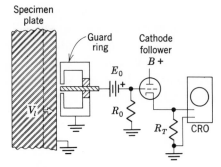

Fig. 7.20 Static capacitor system for measuring rapidly changing surface velocities. (*After Rice.*)

The advent of gas lasers with their very high intensity has led to the development of high-speed interferometers. One measures displacements as a function of time (Barker and Hollenbach, 1965), but the most remarkable one performs differentiation optically to yield velocities directly (Barker, 1967). Its principles are illustrated in Fig. 7.21. The coherent laser beam is directed onto the moving specimen surface by a mirror and then collected and sent

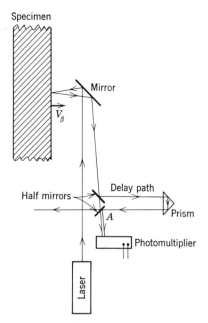

Fig. 7.21 Schematic diagram of velocity-measuring interferometer. (*After Barker.*)

to a half mirror, where it is split into two beams, one of which travels along a delay path. The two beams are recombined at point A, where the direct beam has come from the specimen surface at a time later than the delayed beam. If the specimen surface was at a position x at the earlier time, then its position is $x + \Delta x$ at the later time $t + \Delta t$. Thus the phase relation between the two beams is proportional to the specimen-surface velocity. This method allows velocities as high as about $3.4 \times 10^3 \pm 3.4$ cm/sec to be measured with a time resolution of about 2×10^{-9} sec.

Some representative measured velocity-time curves are shown in Fig. 7.22,

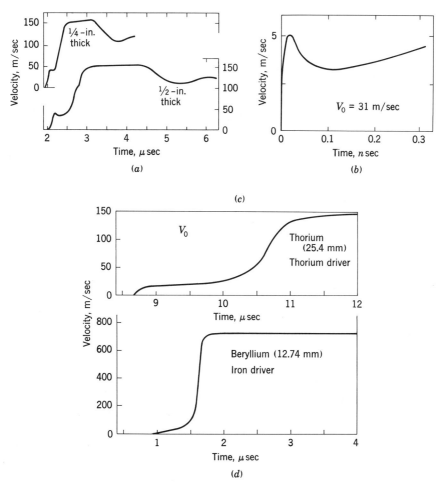

Fig. 7.22 Back-surface velocities of impacted plates as a function of time for some typical materials: (a) two thicknesses of Armco iron (Taylor and Rice); (b) aluminum (Barker, Butcher, and Karnes); (c) thorium (Taylor and Rice); (d) beryllium (Taylor and Rice).

as measured by Taylor and Rice (1963) and by Barker, Butcher, and Karnes (1966). In each case the longitudinal elastic wave arrives at the back surface first and causes the velocity to rise. As the velocity (that is, the stress) rises, plastic flow occurs at an increasing rate so that plastic relaxation decreases the rate of rise (and in the cases of iron and aluminum actually makes the rate become negative). Later the pressure wave arrives, and since the flow cannot relax it, the stress again rises rapidly. Note that the pressure wave takes twice as long to arrive in a $\frac{1}{2}$-in.-thick iron plate as in a $\frac{1}{4}$-in. one; this is consistent with the fact that its velocity is constant.

The effects of alloying and heat-treating on steels are illustrated in Fig. 7.23 with data obtained by Jones, Neilson, and Benedick (1962). Here the

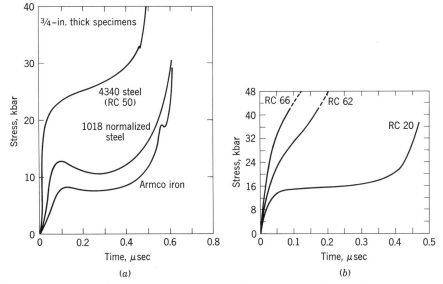

Fig. 7.23 Stress-time curves measured for various steel compositions and heat treatments: (a) effect of carbon and alloy additions on the behavior of steel; (b) high carbon–high chromium tool steel (Hampden) in three initial states: annealed—RC 20; As-quenched—RC 66; quenched and tempered—RC 62.

velocities have been converted into stresses. The impact curves may be seen to have the same general forms as quasi-static stress-strain curves. Also, it is notable that even with high stress levels and rates, dislocation motion is stable in steels, so viscous mechanisms rather than quasi-static pinning mechanisms determine the mechanical behavior.

As the longitudinal disturbance propagates through a plate, its energy gradually diminishes as a result of plastic flow. This may be seen in

Fig. 7.22a, where the early maximum in the velocity (upper yield velocity) is less for the $\frac{1}{2}$-in. plate than for the $\frac{1}{4}$-in. one. Also, the initial rise time is longer for the $\frac{1}{2}$-in. plate. Taylor and Rice (1963) have measured this yield-point-attenuation effect in Armco iron, and some of their results are shown in Fig. 7.24.

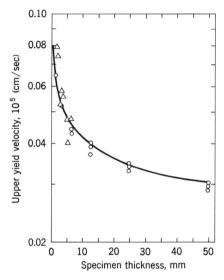

Fig. 7.24 Upper-yield-velocity attenuation curve for Armco iron. (*After Taylor and Rice.*)

The graph uses semilogarithmic coordinates so the attenuation is very rapid in thin plates (or in the frontal parts of thick ones) and then slows down for greater thicknesses. This is quite consistent with dislocation behavior, as will now be shown.

COMPUTED WAVE PROFILES

In order to calculate wave profiles in impacted plates, it is necessary to provide an explicit form for Eq. (7.56), that is for the strain-rate (constitutive) equation. The form used for the computations that will be described here is the one proposed by Gilman (1965) and tested by Taylor (1965) for the case of Armco iron. It is

$$\dot{\epsilon}_p = bv^*(N_0 + M\epsilon_p)e^{-(D+H\epsilon_p)/\sigma_s} \tag{7.59}$$

and the meanings of the constants plus values chosen to represent iron are as follows:

b = Burgers displacement = 2.5×10^{-8} cm
v^* = terminal dislocation velocity = 3.2×10^5 cm/sec
N_0 = initial dislocation density = 10^6 disl/cm^2
M = multiplication constant = 10^{11} disl/cm^2
D = characteristic drag stress = 12.0 kbar
H = strain-hardening coefficient = 4.1 kbar

In addition, the following material properties were used for the equation of motion:

p = mass density = 7.85 g/cm^3
G = shear modulus = 814 kbar
C_l = longitudinal wave velocity = 6.0×10^5 cm/sec

The problem to be studied numerically is that of a plate 0.6 cm thick that moves at a speed of 10^4 cm/sec until it strikes a stationary plate that is twice as thick (1.2 cm). Both plates are made of the same material, so the velocity of the interface between them at the instant of impact becomes $v_0/2$, or 5×10^3 cm/sec.

For a sequence of times following impact the governing equations (7.53), (7.56), (7.57), and (7.58) were solved by means of the HEMP Computing Code developed by M. L. Wilkins. The machine used was a CDC 3600, and the results were plotted out by means of CalComp. for every 0.2 μsec after the time of the impact. The Von Neumann artificial viscosity coefficient was carefully adjusted to minimize the oscillations in the solutions, so the structure in the output curves is real structure in the solution of the equations and not numerical artifact. Some of the results are shown in Fig. 7.25, which presents velocity versus distance profiles for various time intervals after the impact at $t = 0$.

At the instant of impact, the velocity of the material to the left of the zero distance coordinate is 10^4 cm/sec, and zero for the material to the right. After 0.2 μsec have passed, the material to the right acquires some velocity, while the material to the left loses some. Both are thereby put into compression. As the compressive stress (proportional to the velocity) rises, however, it induces an increasing plastic flow rate which relaxes the shear stress and causes the velocity to pass through a maximum and then decrease. After it

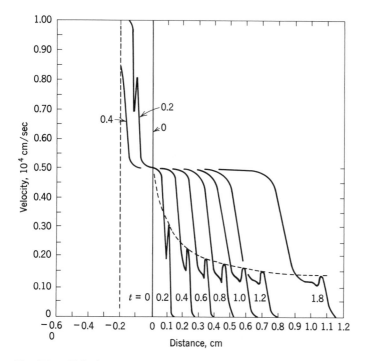

Fig. 7.25 Velocity profile of a mechanical disturbance propagating through an impacted plate, calculated by means of the HEMP code. The time parameter is given in microseconds. The dashed curve indicates how the upper yield velocity attenuates during propagation.

passes through a minimum, the velocity rises again as strain-hardening occurs, and the relatively slower pressure wave reaches the position of interest.

Figure 7.25 shows quite clearly how the longitudinal elastic wave gradually separates from the pressure wave because of its higher velocity. The figure also shows that the profiles calculated from dislocation theory have the same qualitative shapes as the experimental curves of Fig. 7.22. Close quantitative agreement can be obtained by adjusting the material constants within ranges that are consistent with the results of quasi-static measurements.

Another feature of Fig. 7.25 is the strong attenuation of the upper yield velocity (dashed curve). This is in good agreement with the experimental measurements of Taylor and Rice as shown in Fig. 7.24. Quantitative agreement has been demonstrated by Taylor (1965).

The maximum shear-stress profile in a plane wave is of some interest because it indicates how the driving force for plastic flow varies. An example

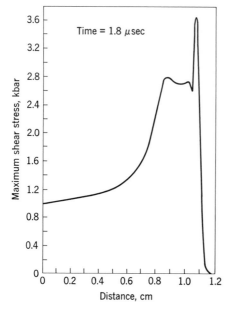

Fig. 7.26 Shear-stress profile of the wave in Fig. 7.25
at 1.8 μsec after impact.

of a computed shear-stress profile is shown in Fig. 7.26. Here the shear stress
rises to a maximum (upper yield point) at the front of the longitudinal elastic
wave. It then follows a low plateau whose height is determined by the rate
of plastic relaxation until the pressure wave appears, and the shear stress
gradually decays.

The effects of varying the material parameters were studied numerically,
and some results are presented in Figs. 7.27 and 7.28. In particular, the effects
of variations in the drag stress and the multiplication coefficient are presented. Figure 7.27 shows profiles at constant time for three drag stresses,
and it is apparent that high values of the drag stress lead to high upper yield
velocities and small amounts of plastic relaxation. Another point of interest
is that a dynamical interaction occurs between the trailing edge of velocity
peak and the leading edge of the plateau, and this leads to a decrease in the
rise time of the pressure part of the disturbance. This effect was first noted
by M. L. Wilkins, and it arises naturally from the microdynamical theory,
so it constitutes an original result of the theory. Figure 7.28 shows that the
multiplication coefficient must exceed some minimum value in order for a
significant amount of plastic relaxation to occur during the short available
time. This is why a substantial rate of cross glide is necessary for good
ductility.

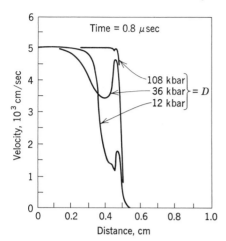

Fig. 7.27 Wave profiles for various drag stresses at the same time after impact of two plates (all other parameters held constant).

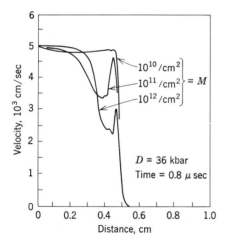

Fig. 7.28 Calculated wave profiles for various dislocation multiplication coefficients at the same time after impact of two plates (all other parameters held constant).

The functional form of Eq. (7.59) implies that the material has a definite yield stress that is determined by the drag stress, but not all materials behave in this way. Some, for example, high-purity metals, exhibit linear viscosity at low stress levels; that is, they begin to flow at a measurable rate as soon as a small stress is applied, and the rate increases in proportion to the stress. Of course, a limit to the rate exists, so the behavior becomes non-linear at high stress levels. A function that can be used to describe this behavior is the hyperbolic tangent with the dislocation velocity written as [this is an alternate to Eq. (7.31)]:

$$v = v^* \tanh \frac{\sigma b}{B v^*} \tag{7.60}$$

where B is the damping coefficient. The effect of using this function in Eq. (7.59), instead of the exponential "yield function," is illustrated by Fig. 7.29. The value used for B was 7×10^{-4} dyn-sec/cm², which is typical for

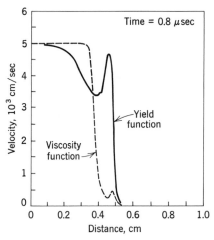

Fig. 7.29 Comparison of computed wave profiles for two different dislocation velocity functions (other parameters held constant).

pure metals. This resulted in very rapid plastic relaxation of the shear stresses, so the rate of rise of the pressure is high once the shears have been relaxed.

REFERENCES

Akulov, N. S.: *Acta Met.*, **12**: 1195 (1964).

Alexander, H., and P. Haasen: *Solid State Phys.*, **22**: 28 (1968).

Barker, L. M.: *IUTAM Symposium*, Paris, September, 1967.

───── and R. E. Hollenbach: *Rev. Sci. Instr.*, **36**: 1617 (1965).

————, B. M. Butcher, and C. H. Karnes: *J. Appl. Phys.*, **37**: 1989 (1966).

Bodner, S. R.: "Mechanical Behavior of Materials under Dynamic Loads," E. S. Lindholm (ed.), p. 176, Springer-Verlag, New York, 1968.

Butler, J. F.: *J. Mech. Phys. Solids*, **10**: 313 (1962).

Campbell, J. D.: "Dislocation Dynamics," Rosenfield et al. (eds.), McGraw-Hill Book Company, New York, 1968.

Cottrell, A. H.: *Proceedings of the Conference on Relation between Structure and Strength in Metals and Alloys*, H. M. Stationery Office, London, 1963.

————: *Report of the Conference on Strength of Solids*, p. 30, Physics Society of London, 1948.

Dey, B. N.: *J. Appl. Phys.*, **38**: 4144 (1967).

Garafalo, F.: "Fundamentals of Creep and Creep Rupture in Metals," The Macmillan Company, New York, 1965.

Gillis, P. P.: *Fifth International Symposium on High Speed Testing*, Boston, March, 1965.

———— and J. J. Gilman: *J. Appl. Phys.*, **36**: 3370 (1965).

Gilman, J. J.: "Dislocation Dynamics," Rosenfield et al. (eds.), McGraw-Hill Book Company, New York, 1968.

————: *J. Appl. Phys.*, **36**: 2772 (1965).

————: *Trans. AIME*, **206**: 1326 (1956).

Govorkov, V. G., V. L. Indenbom, V. S. Papkov, and V. R. Regel: *Sov. Phys.-Solid State*, **6**: 802 (1964).

Haasen, P.: *J. Phys. Radium*, **27**: C3–30 (1966).

Hahn, G. T.: *Acta Met.*, **10**: 727 (1962).

————, C. N. Reid, and A. Gilbert: *Acta Met.*, **10**: 747 (1962).

Hart, E. W.: *Acta Met.*, **3**: 146 (1955).

Hughes, D. S., L. E. Gourley, and M. F. Gourley: *J. Appl. Phys.*, **33**: 3224 (1962).

Ivanov, A. G., S. A. Novikov, and V. A. Sinitsyn: *Sov. Phys.-Solid State*, **5**: 196 (1963).

Johnston, W. G.: *J. Appl. Phys.*, **33**: 2716 (1962).

———— and J. J. Gilman: *J. Appl. Phys.*, **30**: 129 (1959).

Jones, O. E., F. W. Neilson, and W. B. Benedick: *J. Appl. Phys.*, **33**: 3224 (1962).

Kolsky, H.: "Stress Waves in Solids," Clarendon Press, Oxford, 1953; also, Dover Publications, Inc., New York, 1963.

Li, J. C. M.: in "Recrystallization, Grain Growth, and Textures," p. 45, American Society for Metals, Cleveland, Ohio, 1966.

————: *Acta Met.*, **11**: 1269 (1963).

Marsh, S. P., and R. G. McQueen: *Bull. Am. Phys. Soc.*, **50**: 506 (1960).

Minshall, F. S.: *J. Appl. Phys.*, **26**: 463 (1955).

Mura, T.: *Phys. Stat. Solidi*, **10**: 447 (1965).

Patel, J. R., and A. R. Chaudhuri: *J. Appl. Phys.*, **34**: 2788 (1963).

Reppich, B., P. Haasen, and B. Ilschner: *Acta Met.*, **12**: 1283 (1964).

Rice, M. H.: *Rev. Sci. Instr.*, **32**: 449 (1961).

————, R. G. McQueen, and J. M. Walsh: *Solid State Phys.*, **6**: 1 (1958).

Taylor, J. W.: from a private communication of 1968.

————: *J. Appl. Phys.*, **36**: 3146 (1965).

———— and M. H. Rice: *J. Appl. Phys.*, **34**: 364 (1963).

Venable, D.: *Phys. Today*, **17**: 19 (1964).

Webster, G. A.: *Phil. Mag.*, **14**: 775 (1966).

Wilkins, M. L.: Calculation of Elastic-Plastic Flow, *Methods Computational Phys.*, **3**: 211 (1964).

8

FLOW
RESISTANCE

Both analysts who attempt to identify basic mechanisms and synthesists who work on designs for new materials are faced by a colossal collection of facts and ideas when they examine the subject of flow resistance. Furthermore, they both face situations that cannot be fully defined at the present time. The resistive mechanisms can be labeled with names and classified, but most real materials are so complicated in structure that it is difficult to demonstrate the relative importance of a given mechanism in a particular set of circumstances. Therefore, measurements of average responses remain indispensable to the designer, but mechanistic ideas may help to suggest qualitative design changes.

From the dynamical viewpoint, the effect of flow resistance is to reduce the flow rate that is induced by a given applied stress. Either a decrease in the number of participating dislocations or a decrease in their average velocities can reduce the rate of flow. Most of the discussion here will be concerned with the latter effect. Changes in the mobile population have been discussed in Sec. 6.2.

In Sec. 5.3 it was mentioned that dislocations move in a highly dissipative way, with most of the plastic work being converted into heat. Farren and Taylor (1925) found in fact that 95 percent of the plastic work done on aluminum crystals during strains of 15 to 53 percent at room temperature is converted into heat. This has been confirmed by other authors. Furthermore, the 5 percent that is stored is just what is needed to produce the observed intensity of strain-hardening (Nabarro, 1967).

8.1 VISCOUS RESISTANCE

Because so much dissipation occurs, there is no unique way of describing the details of the flow process. Also, a close analogy with liquids in which all the deformational work is dissipated might be expected. Because of this analogy, there is a distinct advantage in applying ideas associated with the behavior of fluids to the dislocation case. In this way, discussions of crystalline and noncrystalline solids can be unified. Also, the highly developed theory of transport in fluids can be applied to the microscopic shear events within crystals. It must be remembered, of course, that permanent structural changes occur during the flow of solids at low temperatures. This is not the case for liquids, so the behavior is definitely different in spite of some similarities.

Since dislocations have both micro- and macro-aspects, the viscous resistance to their motion must be described in two stages. First, as was discussed in a previous section, semi-macromechanics is used to relate the behavior to a local viscosity coefficient. The total viscous drag is obtained by integrating the power loss over the whole velocity-gradient field. Then the molecular mechanisms that determine the local values of the viscosity coefficient are considered. The problem is somewhat simplified by the fact that a dominant portion of the loss occurs in the region at the very core of a dislocation. Therefore, attention can be focused on this region where the molecules actually slide over one another.

As a dislocation moves along its glide plane, the elastic strains at points remote from its center undergo changes. Thus the moving dislocation is surrounded by a strain-rate (or velocity-gradient) field. In addition, at the very center the atoms on the top side of the glide plane slide over those on the bottom side. Thus a velocity gradient exists across the glide plane (Fig. 8.1). For a narrow core, its magnitude can be very large compared with the other velocity gradients (strain rates) in the system. Whenever a velocity gradient exists in a material (gas, liquid, or solid), it tends to become decreased as momentum is transported from the higher-velocity regions to those with lower velocities. The viscosity coefficient measures the efficiency of this transport.

Another source of drag on moving dislocations is the anelastic relaxation that can occur if impurities or other defects are present and can move to cause stress relaxation. This type of drag depends strongly on the dislocation velocity (Schoek and Seeger, 1959) and has a relatively small magnitude. It will not be discussed further here, because it is absent in pure materials and point defects usually cause other effects that are larger in magnitude.

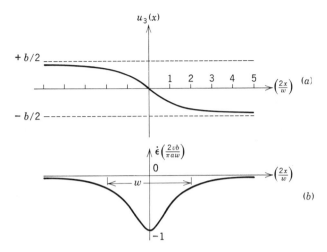

Fig. 8.1 Relation of velocity gradient ($\dot{\epsilon}$) to displacement along the glide plane of a dislocation: (a) displacement function for stationary dislocation; (b) velocity gradient for moving dislocation (neglects perturbations caused by local atomic interactions).

Mason (1960) first emphasized the usefulness of the viscosity viewpoint in considering dislocation losses, and he showed how to calculate the power loss in the strain-rate field. However, he arbitrarily excluded the core region when he integrated over the field. Gilman (1968) showed how the core region can be included in the calculation and that most of the loss occurs there for a given viscosity coefficient. (See Sec. 5.3.)

Because of the localization of the power loss, viscous processes that occur at the glide plane are most important in determining the dislocation damping constant B, which can be written approximately as

$$B \simeq \frac{\eta}{\pi} \frac{b^2}{aw} \tag{8.1}$$

where η is the local viscosity coefficient. This has the fortunate consequence of considerably reducing the number of important mechanisms. Furthermore, if dislocation-line motion is decomposed into a series of kink motions, then the possible sources of viscosity become even more localized.

From the dynamical viewpoint, dislocations provide little or no mechanical advantage in a system unless they reduce the local viscosity coefficient. To illustrate this, consider two solid blocks of material, one sliding over the other on a flat interface where the blocks are separated by a uniform distance a. Suppose that the gap is filled by a medium of viscosity η, with elastic properties similar to those of the blocks. Let the relative velocity between the blocks be v.

In the absence of a dislocation, the drag stress that resists the motion is

$$\sigma_d^0 = \frac{\eta v}{a}$$

Now the drag stress on an edge dislocation of strength b and width w moving at a velocity v_d is

$$\sigma_d \simeq \frac{1}{\pi} \left(\frac{b}{w} \right) \frac{\eta v_d}{a}$$

but

$$v = N b v_d$$

where N is the number of dislocations per unit length along the interface. Thus,

$$\sigma_d^d \simeq \frac{1}{\pi} \left(\frac{1}{wN} \right) \frac{\eta v}{a} \simeq \frac{1}{\pi} \left(\frac{1}{wN} \right) \sigma_d^0$$

so the drag stress in the presence of dislocations is greater than in their absence unless

$$N \gtrless \frac{1}{\pi w}$$

or unless the value of η is changed by their presence.

8.2 SOURCES OF VISCOSITY

In order to smooth out differences in velocities within a medium, momentum must be transferred from regions that are moving fast to slower ones. The means for this to occur were analyzed long ago by Maxwell (1867), and modern developments began with the work of Frenkel (1926) and Andrade (1934). There are two general categories of viscous mechanisms, "gaslike" and "solidlike" (Green, 1952). In the gaslike mode, particles (or quasi-particles) are free to traverse relatively long distances between collisions. As the particles cross an imaginary plane perpendicular to a velocity gradient, they carry more momentum down the gradient (on the average) than they carry up it. Thus the velocity of the slower material tends to increase, while that of the faster tends to decrease. The net velocity relative to some fixed reference tends toward zero.

In the case of solids, the main "gases" are formed by free electrons and phonons. Other excitations that create mobile quasi-particles may also act in this fashion, but their densities may be too small to cause significant viscosity.

In the solidlike mode, direct interactions between sliding molecules tend to smooth out velocity differences. The molecules may be constrained to remain in their own layers, but a faster-moving layer sliding over a slower one exerts a dragging force that tends to speed up the latter. At the same time, the faster layer tends to slow down. On the average, one can think of the sliding molecules as being temporarily coupled together by a force, or per unit area by a coupling shear stress σ_c. If the mean time that the coupling lasts is called τ, then the viscosity coefficient is (Maxwell, 1867)

$$\eta = \sigma_c \tau \tag{8.2}$$

and if a single process causes most of the loss, the damping constant becomes

$$B \simeq \frac{b^2}{aw} \frac{\sigma_c \tau}{\pi} \tag{8.3}$$

The coupling relaxation time may be a function of such factors as the applied stress and the temperature, depending on what particular loss mechanism operates.

For dislocations moving through otherwise-perfect crystals, the Peierls stress couples the molecules across the glide plane. If it is small, as for close-packed glide planes in pure metals, then since τ is roughly the reciprocal Debye frequency (say 10^{-12} sec) and σ_c is certainly less than about 10^5 dyn/cm², the local viscosity coefficient may be as small as $\sim 10^{-7}$ P, which is very small compared with the viscosity of a typical liquid metal ($\sim 10^{-2}$ P).

On the other hand, if the Peierls stress is large, as in covalent crystals such as Ge and Si (say $\sim 10^{10}$ dyn/cm²) and the coupling time is long (say 1 sec per atomic shear event), then the viscosity coefficient is about 10^{18} P, which is very high.

In imperfect crystals, the viscosity is heterogeneous. Dislocations may move quite freely over glide-plane areas that are free of imperfection, impeded only by electron or phonon gas viscosity. At imperfections, strong local bonding may create a strong coupling force across the glide plane, or weak bonding may destroy the local periodicity and thereby raise the effective Peierls stress. Nonviscous drag will also result if the moving dislocation intersects another one and acquires a jog so that it leaves a dipole in its wake. This constitutes a net change in the internal structure and leads to strain-hardening, but it is qualitatively different from a viscous loss mechanism.

8.3 VISCOSITY AT HIGH VELOCITIES

A linear dependence of dislocation velocity on applied stress has been observed in several pure crystals: copper (Greenman, Vreeland, and Wood, 1967); zinc (Pope, Vreeland, and Wood, 1967); germanium (Schafer, 1967). It is apparent, however, that the velocity cannot continue to be proportional to stress indefinitely, because it would soon exceed sonic velocities. Therefore, it is legitimate to wonder how the damping constant depends on velocity. Taylor (1969) has suggested that relativistic effects cause the damping constant to take the form (screw dislocation) [see also, Eq. (7.30)]

$$B = B_0 \left(1 - \frac{v^2}{c_t^2}\right)^{-1} \tag{8.4}$$

where B_0 is the constant at low velocities, v is the instantaneous velocity, and c_t is the transverse elastic wave velocity. According to this expression, the drag force increases without limit as v approaches c_t. Therefore, for any feasible applied force, v cannot exceed c_t (in a linear system).

Taylor's suggestion may be justified in the following simple way. The damping constant has the form $B = p_x/A_y$, where p is momentum and A is area. That is, it measures momentum transfer per unit area. It is well-known that the Lorentz transformation for a momentum component is $= p_x = p_x^0/\beta$, where $\beta = (1 - v^2/c^2)^{\frac{1}{2}}$; and since sliding occurs in one direction, the Fitzgerald contraction of the area is given by $A_y = \beta A_y^0$. Thus Eq. (8.4) follows.

Another approach to the problem of determining the velocity dependence of the viscosity has been presented by Weertman (1969). It is more detailed but less general.

8.4 GASLIKE VISCOSITY

Imagine that a small volume element of material is instantly sheared or compressed a specific amount. This will create a stress in it which the material will attempt to reduce by rearranging its internal structure. If the amount of stress relaxation that can occur is σ_c and the time required for the relaxation is τ, then the viscosity coefficient for the particular relaxation mode is $\eta = \sigma_c \tau$. A moving dislocation is simply a means for successively shearing or compressing the material along its path, so any relaxation mode can contribute viscous resistance to the dislocation's motion. Therefore, the number of distinct mechanisms is large, although many of them contribute only small amounts of viscosity.

Both gaslike and solidlike mechanisms operate. Some of the quasi-particles that can carry momentum down a velocity gradient (especially across the glide plane) are as follows:

1. Phonons; density of this gas increases with temperature
2. Electrons and holes; becomes important at low temperatures where the phonon density is small
3. Excitons
4. Spin waves;
   a. Ferromagnetic
   b. Paramagnetic
   c. Nuclear
5. Local vibrational modes;
   a. Impurities
   b. Internal modes in molecules
6. Local structural rearrangements;
   a. Domain wall motions (electric, magnetic, chemical)
   b. Atomic or molecular positions

The viscosities that arise from several of these have been discussed recently elsewhere (Mason, 1965), so only the most common ones will be considered further here; namely, electron and phonon damping.

As a dislocation (or other acoustic disturbance) moves, its causes the ions of the material to move. This induces electric fields which cause the electrons to move, and since the electrons' rapid motion quickly establishes charge neutrality, they acquire velocities equal to the particle velocity in the material.

Wherever there is a shear displacement rate, so that one layer of material moves faster than another, the electron gas can transport momentum from the faster region to the slower one. For simple viscous behavior, the stress required for steady flow is

$$\sigma_s = \eta \dot{\epsilon}_s \tag{8.5}$$

where the strain rate equals the velocity gradient.

From the kinetic theory of simple gases, the viscosity coefficient is

$$\eta = \frac{Nmv\Lambda}{3} \tag{8.6}$$

where N = number density of particles
$\quad\quad m$ = mass per particle
$\quad\quad v$ = rms velocity
$\quad\quad \Lambda$ = mean free path

An equation of the same form holds for the case of a phonon gas with similar meanings assigned to the individual parameters. However the temperature dependence is different for phonons.

Since the core of a dislocation has a rather definite width w, material in its path becomes stressed at a characteristic rate whose dominant frequency is v_d/w, and this is 10^6 Hz or more for flowing materials. The corresponding wavelength is $\lambda = (v_s/v_d)w$, which is almost always large compared with Λ, so the "gas picture" is a reasonable approximation except at very low temperatures, where Λ becomes large compared with w.

For a simple free-electron gas, only the electrons at the Fermi surface can transport net momentum, and $v^2 = \frac{3}{5}v_f^2$, where v_f is the Fermi velocity. Thus η becomes

$$\eta_e \simeq \frac{Nm\Lambda v_f}{4} \tag{8.7}$$

Values for Λ may be estimated from its role in the theory of resistivity. This allows η to be written in terms of the resistivity ρ (ohm-cm) of the material:

$$\rho_e \simeq \frac{9 \times 10^{11}\hbar^2(3\pi^2N)^{\frac{2}{3}}}{5e^2\rho} \tag{8.8}$$

where $\hbar$ is $1/2\pi$ times Planck's constant and e is the electron's charge. Thus high resistivity means low electron-gas viscosity. Some estimated numerical values for the damping constant B_e are given in Table 8.1.

TABLE 8.1 DAMPING CONSTANTS FOR THE INTERACTION OF DISLOCATIONS AND THE ELECTRON GASES OF SOME METALS

	ELECTRON-GAS DAMPING CONSTANT (10^{-4} P)			
METAL	20°K	50°K	100°K	REFERENCES
Al	37	16	12.0	a
			<1.4	c
Cu	—	20	2.5	a
Pb	8.6	0.10	—	a
		0.86	—	b

a. W. P. Mason, in "Dislocation Dynamics," McGraw-Hill Book Company, New York, p. 487, 1968.
b. A. Hikata and C. Elbaum, *Phys. Rev. Letters*, **18**: 750 (1967).
c. J. A. Gorman, D. S. Wood, and T. Vreeland, *J. Appl. Phys.*, **40**: 833 (1969).

The viscosities of liquid metals at their melting points amount to a few centipoises. The electron-gas viscosities of Table 8.1 are smaller than this, so dislocations in high-purity metals such as aluminum are accordingly very mobile at low temperatures (where the electron-gas viscosity provides most of the resistance to motion).

Instead of a collection of electrons, the gas of Eq. (8.6) may be constituted of phonons, especially if the temperature is high. Just as in the electron case, Mason (1960) has suggested that values for Λ may be estimated from its role in the theory of resistivity, in this case, thermal resistivity. The relation for shear waves is

$$\eta_p = D \left(\frac{E_0 K}{c_v v^2} \right) \tag{8.9}$$

where c_v = specific heat at constant volume

K = thermal conductivity

E_0 = thermal energy density

v = average Debye velocity

D = coupling parameter calculated from nonlinear elastic behavior

Some numerical values for room temperature are given in Table 8.2, and

TABLE 8.2 SOME VALUES OF PHONON–GAS VISCOSITIES FOR SALT CRYSTALS AT ROOM TEMPERATURE [after W. P. Mason, J. Appl. Phys., 35:2779 (1964)]

SALT	$\dfrac{E_0 C_v}{K \bar{V}^2}$	D	η_p
LiF	0.98×10^{-3}	16.1	15.7×10^{-3}
NaCl	1.6×10^{-3}	2.1	3.31×10^{-3}
KCl	2.4×10^{-3}	2.0	4.71×10^{-3}
KBr	3.4×10^{-3}	2.2	7.6×10^{-3}

these values are comparable with the viscosities of liquid metals at their melting points.

At low temperatures, η_p is negligible, and it never becomes larger than about 3×10^{-4} P, although it does appear to limit dislocation mobility in some cases (Pope and Vreeland, 1967). At higher temperatures, it should increase with the temperature as the thermal energy density increases. This has been observed in Al according to Gorman, Wood, and Vreeland (1968).

8.5 SLIDING VISCOSITY (SOLIDLIKE)

It was shown in Sec. 5.3 that most of the viscous loss associated with dislocation motion occurs at the very core because the maximum. velocity gradient is there. Also, at the core, the atoms slide over one another while they are strongly interacting. Various types of interaction occur across the glide plane. Some of them are:

1. Pure-crystal matrix (Frenkel-Kontorova model): interaction across the glide plane causes drag forces:
2. Point defects at glide plane:
   - *a.* Vacancies
   - *b.* Interstitials
   - *c.* Impurities
   - *d.* Trapped electrons (holes) } Cause local drag forces.
   - *e.* Isotopes
   - *f.* Excited atoms
3. Intersecting dislocations.
4. Domain boundaries:
   - *a.* Stacking faults.
   - *b.* Antiphase boundaries.
   - *c.* Bloch walls (magnetic).
   - *d.* Electric domain boundaries.
   - *e.* Crystallite boundaries.
   - *f.* Interfaces between phases.

In addition to these direct modes of power loss, there are some indirect possibilities. One is the production of acoustic radiation either by oscillations of kinks (Eshelby, 1962), or by point defects that lie nearby the glide plane and are excited into vibration by "collisions" with moving dislocations (Takamura and Morimoto, 1963). Another consists of various anelastic relaxations in which the strains around dislocations induce atomic movements (Snoek and Zener relaxations), adiabatic temperature changes, magnetization changes, reversible chemical reactions, or electric polarization changes. As a dislocation moves past a point, the material there receives a cycle of shear strain (or shear plus dilatation). The rate depends on the dislocation velocity, and the amplitude is inversely proportional to the distance from the dislocation. If the stress and strain remain in phase, there is no power loss; but if they get out of phase, loss occurs in proportion to the tangent of the phase angle.

In soft crystals all these mechanisms can cause measurable losses if they are present, but the greatest losses occur as a result of the sliding action at the core. This causes strong interactions between the atoms that lie just above and below the glide plane. For example, in a steel with a yield stress of 10^5 psi $= 6 \times 10^{10}$ dyn/cm^2, $b = 3A$, and a dislocation density of 10^{10}/cm^2, the effective value of η is then $\sim 2 \times 10^7$ P. This may be compared with a typical room-temperature value for the phonon-gas viscosity, $\sim 10^{-3}$ P.

The momentum that is transferred across the glide plane of a moving dislocation depends on the local force and the length of time it acts, because

$$dp = f \, dt \tag{8.10}$$

It was first pointed out by Maxwell (1867) and discussed by Mason (1965, vol. 4, part A, p. 302) that this means the viscosity coefficient may be written as the product of a coupling stress and a local relaxation time [Eq. (8.2)]. Therefore, it is necessary to determine how these factors depend on the local atomic forces, on the applied stress, and on temperature. An exhaustive study would not be appropriate here, but some of the features will be outlined.

Before the effects of various defects can be considered, it is necessary to learn how the coupling stress depends on the atomic structure of a perfectly periodic crystal. This is done by determining how much stress is needed to move a dislocation through a periodic arrangement of atoms. This force can be shown to be quite small with the aid of Fig. 8.2, which is necessarily exaggerated in order to be able to indicate any force at all. Figure 8.2a shows an unlikely crystal structure with an abnormally small ratio of the glide-plane spacing to the Burgers displacement. In Fig. 8.2b an edge dislocation has been introduced with its extra half plane symmetrically disposed with respect to the bottom half of the structure. Because all the forces to the left are balanced by equal and opposite ones at the right, there is no net force on the dislocation.

The configuration of Fig. 8.2c is reached by moving the dislocation from (b). It is also symmetric, so there is no net force on the dislocation. Between the configurations of Fig. 8.2b and c the dislocation is displaced by $b/2$. An additional displacement of $b/4$ leads to the configuration of Fig. 8.2d. Now there is some asymmetry, and the forces are slightly unbalanced, so an applied force is needed to maintain the configuration. Note however, that the force imbalance is quite small.

At Fig. 8.2e the effect of a point defect in creating an asymmetric configuration is indicated.

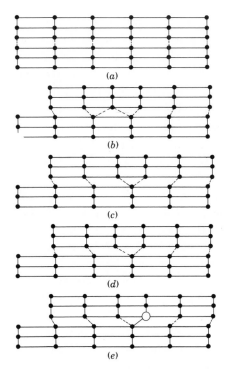

Fig. 8.2 Schematic drawing that indicates how the forces on a dislocation change as it moves.

Since the net force on a moving dislocation is the resultant of adding many larger forces together, an accurate theoretical determination of it is prohibitively difficult at present. The interatomic forces can only be estimated, and this is not good enough. The difficulties can be most clearly stated by describing some results that have been obtained by Kratochvil and Indenbom (1963) and reviewed by Kratochvil (1966) for a one-dimensional dislocation.

The model is known as the Frenkel-Kontorova model (Kontorova and Frenkel, 1938). The purpose is to model the shear and tensile behavior of a crystal simultaneously, and this is done by using springs to give the crystal resistance to tension and a substrate potential to provide shear resistance. These are indicated schematically in Fig. 8.3. The drawing at Fig. 8.3a shows a set of mass points strung together by springs whose stiffness constants are μ; that is, a displacement of one end of one of the springs by an amount x produces a force $= \mu x$. The mass points are also acted upon by a substrate poten-

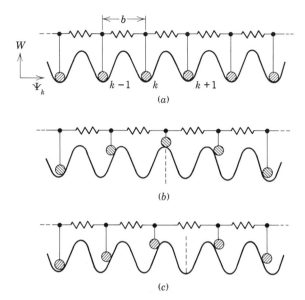

Fig. 8.3   One-dimensional crystal model: (a) perfect
crystal; (b) crystal with symmetric dislocation; (c)
dislocation center moved to second symmetry position.

tial that is indicated schematically in the drawing by a periodic set of "hills"
(the period is the Burgers displacement b).

Let Ψ_k be the displacement of the kth particle in the string (measured
relative to some starting position that is chosen for convenience). Then the
energy of the kth particle depends on the substrate potential-energy function
$W(\Psi_k)$ and on its position relative to its nearest neighbors (longer-range
interactions are not considered). The elastic energy in the spring that lies
between particles k and $k + 1$ is $\mu(\Psi_{k+1} - \Psi_k)^2/2$, since this is the work
required to stretch it. The total energy U of the "crystal" is the sum of the
individual spring and substrate energies:

$$U = \frac{\mu}{2} \sum_{-\infty}^{+\infty} (\Psi_{k+1} - \Psi_k)^2 + \sum_{-\infty}^{+\infty} W(\Psi_k) \tag{8.11}$$

In order for the crystal to achieve static equilibrium, the displacements Ψ_k
must have values such that

$$\frac{\partial U}{\partial \Psi_k} = 0 \qquad \text{at static equilibrium} \tag{8.12}$$

If suitable boundary conditions are applied to the crystal, a dislocation can be put into it as at Fig. 8.3b and c. The required conditions are that the displacement be zero at one end and b at the other:

$$\lim_{k \to -\infty} \Psi_k = 0 \qquad \lim_{k \to +\infty} \Psi_k = b \tag{8.13}$$

Also, it is necessary that the displacements be ordered

$$\Psi_k \le \Psi_{k+1} \tag{8.14}$$

to ensure that only one dislocation is present and not a series with alternating signs.

The equilibrium condition [Eq. (8.12)] produces a set of nonlinear difference equations:

$$\mu(\Psi_{k+1} + \Psi_{k-1} - 2\Psi_k) = \frac{\partial W}{\partial \Psi_k} \tag{8.15}$$

The dislocation that is described by these relations is not free to move in general, because its energy changes somewhat with its position. Thus it has a preferred position in the crystal, and in order to move out of it, the dislocation must surmount a potential-energy barrier with the aid of an applied stress. The effect of the applied stress is included in the theory by modifying the equilibrium equation (8.12) to read

$$\frac{\partial U}{\partial \Psi_k} = F \tag{8.16}$$

where F is the force on each atom caused by the external stress σ_s. Also, the boundary conditions must be modified to allow for the elastic displacements caused by the stress.

An exact solution of the problem has been found for the special case of a parabolic substrate potential as sketched in Fig. 8.4a. This gives the linear dependence of the substrate force on displacement as in Fig. 8.4b. The potential curve is pieced together from the following parabolas:

Region I: $W = C_1\Psi_k^2$

Region II: $W = C_2\left(\Psi_k - \dfrac{b}{2}\right)^2 + C_1\Psi_c^2 + C_2\left(\dfrac{b}{2} - \Psi_c\right)^2$

Region III: $W = C_1(\Psi_k - b)^2$

where C_1 is chosen to give the elastic shear resistance of a perfect crystal for small strains

$$C_1 = \frac{Gb}{2}$$

A parameter m is chosen to determine the maximum shear stress that the

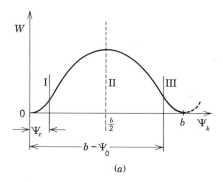

(a)

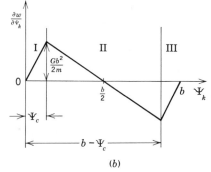

(b)

Fig. 8.4 Substrate potential and force in Kratochvil-Indenbom model: (a) piecewise parabolic potential; (b) linear forces.

perfect crystal will support, which is

$$\sigma_{\max} = \frac{1}{b^2} \left(\frac{\partial W}{\partial \Psi_k}\right)_{\max} = \frac{G}{2m} \tag{8.17}$$

so the greater is m, the smaller is the shear strength. Finally, C_2 and Ψ_c are chosen to make $\partial W/\partial \Psi_k$ continuous at $\Psi_k = \Psi_c$. Their values are

$$C_2 = \frac{Gb}{2(m - 1)}$$

$$\Psi_c = \frac{b}{2m}$$

and the value of the spring constant expressed in terms of the elastic constants is

$$\mu = \frac{E}{b(1 - \nu^2)}$$

where E = Young's modulus and ν = Poisson's ratio.

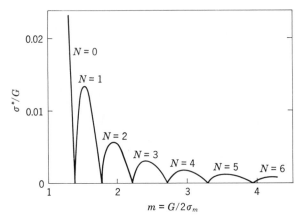

Fig. 8.5 Dependence of critical stress for dislocation motion on the parameter m that determines the shear strength σ_m. N is the number of atoms whose displacements lie in the middle section (II) of the substrate potential.

Results of a calculation of the maximum stress that the dislocation will support are shown in Fig. 8.5, which shows the dependence of the critical stress σ^* on the parameter m that determines the magnitude of the shear strength σ_m. The critical stress is always relatively small and decreases rapidly as the shear strength decreases (m increases). Perhaps the most interesting feature is the fact that the critical stress becomes zero for certain values of m. Notice also that as the width of the dislocation increases (as measured by the number of contained atoms N), the critical shear stress decreases exponentially. These facts, together with the distinctive structure of the curve, indicate that the mobility of a dislocation depends very sensitively on the details of the atomic structure of a crystal. Therefore, the mobility cannot be estimated by means of approximate calculations, and it appears at present that experimental measurements are the only reliable source of information.

It should be mentioned that the above calculation is based on a possible, but unlikely, substrate potential. Therefore, the results are suggestive but have little quantitative meaning. Also, the potential was chosen to be continuous, which is realistic for metals and salts but is probably not a valid condition for covalently bonded crystals (Suzuki, 1963). At displacements where the slope of the potential suddenly changes, the force becomes double-valued, which imbalances the configuration and may result in a large critical stress.

Sanders (1965) has shown with the use of a two-dimensional atomic model that the force which resists the motion of a kink is exceedingly small for a linear substrate potential. It is only about 1 percent of the force needed to move an entire dislocation line. This emphasizes once more the difficulty of calculating forces on dislocations from atomic models.

If the relaxation time is taken to be the atomic vibration period ($\sim 10^{-13}$ sec) and the critical stress is $3 \times 10^{-3}G$ with a modulus of 5×10^{11}, the local viscosity coefficient would be $\sim 1.5 \times 10^{-4}$ P, which is so small that other effects would dominate it.

In imperfect crystals, coupling forces arise when impurities or vacancies lie in the path of a moving dislocation. They destroy the simple periodicity of the material, which introduces local forces of much greater magnitude than those in perfect material.

8.6 NONLINEAR STRESS ACTIVATION

When localized coupling forces exist across glide planes and the temperature is low (compared with the Debye temperature associated with the local coupling force), the flow velocity is not proportional to the applied stress. Instead, it is observed that the flow rate is very small until some critical stress is reached, and then "yielding" occurs. In other words, the flow is stress-activated. In terms of Eq. (8.2), the mean coupling time is a function of stress and rather suddenly decreases when a critical applied stress is reached. A simple analytic form that relates velocity and stress in this case has been proposed by the author (Gilman, 1960):

$$v = v^* e^{-D/\sigma_s} \tag{8.18}$$

where v^* is a terminal velocity and D is a drag stress. This form is followed by several sets of experimental data (Gilman, 1965) and is consistent with the idea that stress can activate dislocation motion via quantum-mechanical tunneling (Gilman, 1968).

At steady state, the work done on a moving dislocation equals the power dissipated in the form of both heat and structural defects. Assuming that heat production dominates, the effective viscous damping constant is the ratio of the driving force per unit length to the velocity:

$$B_{\text{eff}} = \frac{\sigma_s b}{v} = \frac{b}{v^*} \sigma_s e^{D/\sigma_s} \tag{8.19}$$

Thus for the nonlinear case, the effective damping constant is infinite when the stress is zero (the dislocation is "pinned"). Then the damping decreases rapidly with increasing stress until a minimum value is reached when $\sigma_s = D/2$, and finally, it increases nearly linearly with further increases of the stress. The corresponding velocity-stress curve for this solidlike viscosity is compared with the curve for the gaslike viscosity in Fig. 7.14 of the previous chapter.

If stress-activated motion is to be interpreted in terms of a tunneling process, localized kinks or pinning points must be involved, because tunneling can only occur through relatively thin barriers. Locally impeded dislocations can, however, escape their impediments more rapidly via tunneling than by means of thermal activation if the temperature is low and the stress is high.

In the regime where tunneling is important, it may be phonon-assisted, so the rate need not be independent of temperature except at the lowest temperatures.

Dislocation tunneling was first considered by Glen (1956) and Mott (1956), who studied the case of moving dislocations tunneling through "forest" dislocations that intersect them. Their purpose was to account for creep in cadmium and other metals in the liquid helium temperature range. Weertman (1958) has briefly also considered tunneling through a Peierls potential. Gilman (1966) suggested that dislocation tunneling of kinks in semiconductors can be considered in terms of an equivalent electronic process. His model yielded semiquantitative results for the hardness of semiconductors in terms of the energy gap.

Consider a dislocation that interacts locally with a point defect, or a chemical bond in the case of a kink. A quantum-mechanical treatment of the interaction would be quite complicated, so we shall be satisfied with a model potential whose form is consistent with atomic theory. Such a potential is

$$U(x) = -U_0(1 + |x|)e^{-|x|} \qquad (8.20)$$

where $x = z/\delta$ = reduced displacement of the center of the dislocation, with δ being the "range" of the potential, and z the displacement. This potential is simple and yet appropriate, because it yields a linear restoring force for small displacements and decays like a wave function for large x. Also, it is symmetric about $x = 0$. The total binding energy per unit length is U_0, and the binding may be a consequence of both "elastic" and "chemical" interactions.

The restoring force that acts against a relative displacement is

$$f = -\frac{\partial U}{\partial x} = -U_0|x|e^{-|x|} \qquad (8.21)$$

and this reaches its maximum magnitude when $x = 1$ or $z = \delta$. For a short-range interaction, the force cannot extend beyond about one atomic distance, so $\delta \simeq b/2$.

An estimate of the force constant can be made on the basis of Cottrell's elastic-interaction theory (Cottrell, 1953). From this, $U_0/\delta = 4Gb\epsilon$, where ϵ is the mismatch strain of a point defect.

In covalent crystals, U_0 may also be estimated in terms of bond energies.

In Fig. 8.6 the binding potential is sketched, together with the driving

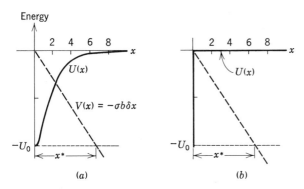

(a)   (b)

Fig. 8.6   Energy-configuration diagrams for a short segment of a bound dislocation: (a) curved potential well;
(b) square-well approximation.

potential that results from the applied stress. The force caused by the applied stress σ_s is $\sigma_s b^2$, so the potential energy associated with a displacement of the dislocation segment is

$$V(x) = -\sigma_s b \delta x \tag{8.22}$$

and the total energy of the system is

$$U(x) = -U_0(1 + |x|)e^{-|x|} - \sigma_s b \delta |x| \tag{8.23}$$

The sketch also shows that two positions exist which have the same energy. Thus the dislocation segment can tunnel from $x = 0$ to $x = x^*$, and the rate can be appreciable if x^* becomes small enough. At $x = x^*$ it may be seen that $U_0 \simeq \sigma_s b \delta x^*$, so

$$x^* \cong \frac{U_0}{b\delta\sigma_s} \tag{8.24}$$

Since the tunneling distance depends inversely on the local stress, high stresses can cause it to be small.

Figure 8.6b shows how $U(x)$ can be approximated by a square-well potential. This is a standard procedure in the theory of the field emission of electrons and leads to a well-known expression for the tunneling probability. However, it does not take into account the change in the barrier height caused by $V(x)$. A somewhat better approximation is one that Zener (1934) developed for calculating the rate of tunneling from one energy level to another. Except for a numerical factor, both procedures lead to the same result for the probability:

$$\exp\left(\frac{-mU_0^2}{4\hbar^2\sigma_s b}\right) \tag{8.25}$$

This can be put into a simple form by using some approximate relations. First, the proportionality $U_0 \sim Gb^2$ and the effective-mass approximation $m \simeq \rho b^3$. Then, the recognition that the frequency of the local oscillator is $\simeq (G/\rho b)^{\frac{1}{2}}$, so an identification between frequency and energy can be made; $\hbar\omega \simeq Gb^3$. Substituting these into Eq. (8.25) then yields the following for the probability of motion:

$$P_m = e^{-\beta G/\sigma_s} \tag{8.26}$$

where β is a numerical coefficient that is difficult to calculate from theory alone, but which can be found from experiments.

Equation (8.25) indicates that the tunneling probability, and hence the rate of stress-activated flow, is a very strong function of the applied stress. Also, the stress required to cause a given tunneling rate is proportional to the local shear modulus (which measures the binding energy), according to Eq. (8.26).

It has been shown elsewhere (Gilman, 1965) that the stress dependence of dislocation velocities in several substances has the functional form of Eq. (8.26). This includes LiF, CaF_2, NaCl, W, Ni, Nb, Fe (3 percent Si). In all these cases, the velocity is controlled by interactions with impurities which cause the localized pinning.

Proportionality of flow stress and shear modulus at low temperatures, as predicted by Eq. (8.26), is shown in Figs. 8.7 and 8.8. The case of tetrahedrally bonded crystals is presented in Fig. 8.7, and that of pure f.c.c. metals in Fig. 8.8. In the former case it is the interatomic bonds that pin the dislocations, while dislocation intersections play this role in the latter case.

8.7 EFFECTS OF TEMPERATURE

The rate of escape of dislocations from bound states depends on the product of the attempt frequency and the success probability. The attempt frequency

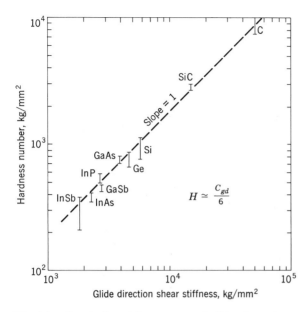

Fig. 8.7   Correlation of flow stress and glide-plane shear stiffness for tetrahedrally bonded crystals.

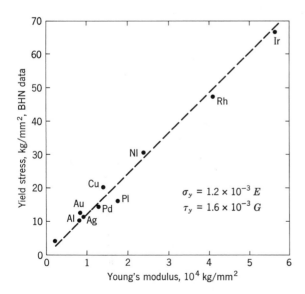

Fig. 8.8   Proportionality of flow stress and elastic stiffness for f.c.c. metals at low temperatures.

will approximately equal the Debye cutoff frequency, or $\sim 5 \times 10^{12}$ Hz, even at very low temperatures, because there will be zero-point motion in the local oscillator. Therefore, the temperature dependence of the rate is not sensitive to this factor, and it is concluded that the important factor is the temperature dependence of the coefficient in the exponent of Eq. (8.25), that is, U_0.

It seems unlikely that a single model for the temperature dependence will suffice, because there are at least two rather different limiting cases, namely, those of "tight" and "loose" binding. In the tight-binding case, the binding potential well is relatively deep, and the main effect of temperature is to raise the average kinetic-energy level at which the dislocation attempts to escape. This will be called *phonon-assisted tunneling* here.

In the case of loose-binding, the dislocation is bound at a particular site, because the glide plane is aperiodic at this site. This causes the dislocation's core to shrink in width enough to reduce its mobility. As the temperature increases, however, the uncorrelated thermal motions cause the width to increase and, therefore, the mobility. At still-higher temperatures, it is more appropriate to describe the process as being thermally activated. In this regime the thermal energy kT plays the primary role in setting the success probability, instead of the stress energy σb^3. The various effects of temperature are described in somewhat more detail below.

For the purpose of predicting the overall response of a solid to a new set of loading conditions, the influence of temperature may be taken into account simply by assuming that the drag stress D has the same variation with temperature as the yield stress. This will yield a good first approximation to the behavior, but says nothing about the mechanism of the temperature dependence.

PHONON-ASSISTED TUNNELING

At high temperatures, flow processes occur as a result of thermal activation that is biased by relatively small applied stresses. In the low-temperature case that is under discussion here, the flow is primarily activated by the stress, but the tunneling rate can be enhanced by thermal-energy input. This leads to a temperature dependence different from the type obtained when thermal activation dominates.

In the tight-binding case, the effect of temperature is to change the effective-energy barrier seen by a bound dislocation. Perhaps this is most clear in the case of a dislocation bound to an impurity. In the absence of thermal energy (and neglecting the zero-point energy), the dislocation segment must tunnel unassisted away from the stationary impurity. But at finite temperatures the impurity vibrates, and part of the time it moves away from the dislocation, thereby tending to assist its escape. This can be expressed in terms of a change in the average energy of the initial state of the system.

Since the binding that is considered here is atomically localized (whether it be caused by an impurity, point defect, intersecting dislocation, etc.), the binding site is expected to behave like an Einstein oscillator (approximately). Therefore, if its frequency is ω, its average energy will be

$$\langle u \rangle = \frac{\hbar\omega}{2} \coth \frac{\hbar\omega}{kT} \tag{8.27}$$

and the energy barrier will be reduced by this amount, yielding

$$U(T) = U_0 - \langle u \rangle \tag{8.28}$$

For a given rate of escape from the barrier (which also means a given dislocation velocity, or a given flow rate at constant dislocation flux), the value of the exponent in Eq. (8.26) must take a particular value. Therefore, if σ_0 is the stress needed to give a certain rate when $T = 0$, some algebraic manipulation shows that the stress needed at a higher temperature is

$$\sigma_s = \sigma_s^0 \left(1 - \frac{\langle \omega \rangle}{U_0}\right) \tag{8.29}$$

or, since $\hbar\omega$ is expected to be proportional to U_0

$$\frac{\sigma_s}{\sigma_s^0} \cong 1 - \xi \coth \frac{\hbar\omega}{kT} \tag{8.30}$$

where ξ is a constant which depends on the ratio of the oscillator energy to the binding energy. This function is plotted in Fig. 8.9 for a few values of ξ in order to show its form. According to the figure, the stress decreases only slowly with increasing temperature at low temperatures, and this is in qualitative agreement with experimental hardness measurements as shown in Fig. 8.10. The larger the ratio ξ, the more effective the thermal excitations as the barrier becomes thinner. No direct measurements are available for comparison with the theory.

A more dramatic effect is the dependence on temperature of the rate of escape at constant stress. This dependence takes the form

$$e^{B \tanh \frac{T}{\theta}} \tag{8.31}$$

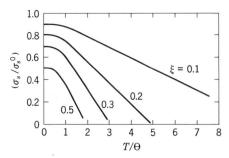

Fig. 8.9 Graph of Eq. (8.30), showing the dependence of the reduced stress on temperature for various values of the ratio of oscillator to binding energy; $\theta = k/\hbar\omega$.

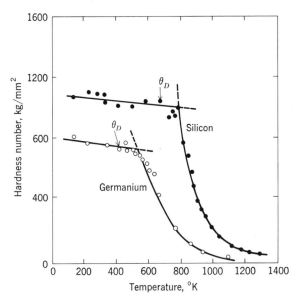

Fig. 8.10 Effect of temperature on the hardnesses of Ge and Si. (*After Nikitenko, 1967.*)

where B and Θ are constants. Thus, for small T/Θ, the escape rate is proportional to

$$e^{cT} \tag{8.32}$$

so it increases rapidly as the temperature increases. This has actually been observed between 75 and 100°K for pyramidal glide in zinc by Blish and Vreeland (1968).

CHANGE OF DISLOCATION WIDTH

General theory as well as experiments with vibrating bubble rafts both lead to the conclusion that the width of a dislocation increases as the temperature increases (Nabarro, 1967, p. 176). If the flow-stress criterion of Sec. 8.5 (Fig. 8.5) is used, the flow stress will have the form

$$\sigma_s = \sigma_s^0 e^{-2\pi w/b} \tag{8.33}$$

where σ_s^0 is the stress at $0°K$ and w is the core width. In accordance with the study of Kuhlmann-Wilsdorf (1960), the width should vary with temperature as

$$w = w_0[e^{\Theta_E/T}(e^{\Theta_E/T} - 1)^{-1}] \tag{8.34}$$

where w_0 is the low temperature width and Θ_E is the Einstein temperature of the local mode. For $T < \Theta_E$ this gives relatively little temperature dependence, but for $T > \Theta_E$ the width becomes proportional to the temperature, so the variation of the flow stress with temperature becomes

$$\sigma_s = \sigma_s^0 e^{-\mu T} \tag{8.35}$$

which is the variation that is observed for LiF (Fig. 8.11) and a variety of other materials. The coefficient μ is a constant of the system (Johnston, 1962).

BEHAVIOR AT HIGH TEMPERATURES

It is well-known that materials soften as their temperatures are raised, but this softening has such a variety of causes that there is not space to discuss it in detail here. In heterogeneous materials, the boundaries between grains or phases are often the first elements that soften. In more homogeneous ones, softening often begins when the structural damage that causes strain-hardening begins to "dissolve." If structural damage is produced in proportion to the plastic strain and the resulting stress increment is proportional to the damage, then the hardening rate is $H = (\partial\sigma/\partial\epsilon)\sigma$. But at the same time, diffusion processes may allow the damage to dissolve and thereby relax the stress in proportion to time. Thus there may be a softening rate $S = -(\partial\sigma/\partial t)\epsilon$. At steady state, the resulting flow rate will be

$$\frac{d\epsilon_p}{dt} = -\frac{S}{H} \tag{8.36}$$

Only in relatively rare cases of pure covalently bonded crystals does the softening result from a change in the mode of activation of the gliding of individual dislocations. Examples are germanium, silicon, aluminum oxide,

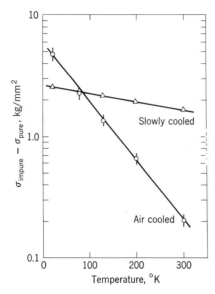

Fig. 8.11 Temperature dependence of the impurity contribution to the flow stress of LiF. Crystals were initially from 300°C at rates of 0.002°C/min and 50°C/min, respectively. (*After Johnston,* 1962.)

prismatic glide in zinc, titanium carbide, and similar "hard" glide systems. The case that has been studied most vigorously is that of germanium, and a careful review of the studies has recently been prepared by Alexander and Haasen (1968).

8.8 STRUCTURAL DEBRIS

Figure 4.21 of the discussion of dislocation geometry illustrated one of the important processes that leads to disturbed glide planes in the wakes of moving dislocations. Numerous jogs may exist along dislocation lines, and motion of the line then causes dipoles to stream out behind it. This has been confirmed experimentally by Dash (1958) for Si; Johnston and Gilman for LiF (1960); Fourie and Murphy for Cu (1962).

DIPOLE TRAILS

If the dipole energy per unit length is S_1 and the jog concentration is C_j, the energy required for a small advance dx of a jogged dislocation is $S_1 C_j \, dx$. The work done by the applied stress field during the advance is $\sigma_s b \, dx$. Thus the incremental stress associated with this mechanism is

$$\sigma_{s1} = \frac{C_j S_1}{b} \tag{8.37}$$

and this can be substantial if C_j is moderately large, say $10^5/\text{cm}$, and S_1 is 10^{-3} ergs/cm. The value of σ_{s1} becomes about 3×10^8 dyn/cm². At the very least this mechanism accounts for the smaller mobilities of screw dislocations relative to those of edge ones.

The stabilities of dipoles are limited, and if the shear stress that acts on a dipole of height h exceeds the value (Cottrell, 1953, p. 152)

$$\sigma^* = \frac{Gb}{8\pi(1 - \nu)} \frac{1}{h} \tag{8.38}$$

the dipole decomposes into monopoles. Here G is the shear modulus, b is Burgers displacement, and ν is Poisson's ratio. This is one reason why the attrition coefficient depends inversely on the stress, Eq. (6.19).

PLANAR FAULTS

In a crystal with an ordered structure, the passage of a dislocation through it may change the degree of order on the glide plane leaving a structural fault in the dislocation's wake. A corresponding increase in energy occurs which creates an opposition to the motion of the dislocation. The order may be of the chemical, magnetic, or electric type. If there is full ordering, no structural fault will be left behind a "superdislocation" whose Burgers vector is a unit translation vector of the superlattice. A partial dislocation will leave a wake, however. Also, if the structure has short-range order (or disorder) then the passage of a dislocation will change the state of order by changing the components of pairs across the glide plane.

If the planar structural fault in the wake of a dislocation has an energy per unit area S_2, then the incremental stress required for its production will be

$$\sigma_{s2} = \frac{S_2}{b} \tag{8.39}$$

This can be small or large depending on the severity of the fault. If the fault is severe with an energy of 10^3 ergs/cm², then the incremental stress will be large, about 3×10^{10} dyn/cm².

REACTION PRODUCTS

The products of reactions like those described in Sec. 4.14 are metastable and, therefore, remain at the places where they have formed. They interact through their stress fields with subsequent dislocations that approach their vicinities. Thus the motions of subsequent dislocations may be impeded.

8.9 ATTRITION

Since a given plastic strain rate is associated with a certain number of mobile dislocations that move at some average velocity, any change in the number of participating dislocations will require the remainder to move faster in order to maintain the given strain rate. But the average velocity can increase only if the average stress increases, so this effect contributes a "resistance to flow."

The primary cause of attrition in the mobile dislocation population is interactions that lead to annihilation. This happens when two parallel edge dislocations of opposite sign meet or when two screw dislocations of opposite handedness encounter one another (Sec. 4.14).

Surface egress is another source of attrition, especially in very thin pieces of material. A special case of this occurs in polycrystals when dislocations move out of grains into grain boundaries.

8.10 COLLECTIVE BEHAVIOR

The dislocations inside a flowing crystal are often spaced so close together, and therefore interact so much, that they cannot be considered as a set of independent entities. Instead, their collective behavior must be considered. It is difficult to identify the effects of collective behavior when the plastic flow is steady (or nearly so). However, transient effects caused by various kinds of perturbations can be observed. It is believed that some of these can be identified as results of many-bodied scattering interactions.

Since the effective masses are usually small compared with elastic forces that act on them, acceleration effects are not important for individual dislocations. However, when they behave collectively, the effective mass of a group can be large in comparison to the coupling forces within the group. Then it is necessary to consider the strain acceleration equation

$$\ddot{\epsilon} = b(N\dot{v} + v\dot{N}) \tag{8.40}$$

This equation describes what happens when the conditions of a flowing material are suddenly changed.

In many transient situations, the second term in the equation (which gives the strain acceleration caused by changes of the dislocation density) is the dominant one. However, the first term provides the possibility of purely accelerative effects when the total line length is constant. One experimental observation that appears to correspond with this is that of transient upper yield points. Sometimes, when a flowing specimen is temporarily unloaded and then reloaded, the stress rises transiently above the previous flow curve, but then returns asymptotically to the projection of the original flow curve. This latter observation is important, because it indicates that the stress overshoot is a true dynamical transient rather than a result of changes of $\dot{N}$. If the term in $\dot{N}$ were the important one, the stress would not be expected to return to the extension of the original flow curve but, instead, to a level either above or below it.

Another factor is that the negative slope of the second half of the stress transient reflects unstable flow. This in turn implies the propagation of a plastic front of some kind. Such a wavelike propagation is more likely to be associated with acceleration effects than with dislocation density changes in a crystal that has already undergone substantial plastic strain.

Consider a set of dislocations that has stopped moving because the stress has been removed from the specimen it is in. When the stress is replaced, the whole set will not move in synchronism, because this mode of motion is unstable. This is analogous with a line of cars for which a traffic light suddenly turns green. Instead, a pulse of motion will begin at the front of the set and propagate back into it. If the damping forces that act on the dislocations are small, the pulse will propagate at nearly the speed of sound. But if substantial damping is present, the response will be slower, and this will tend to make the stress overshoot in a constant displacement rate test.

Prigogine's (1961) discussions of the statistical dynamics of traffic flow also suggest some interesting collective effects for dislocations. Let the dislocation velocity distribution function be $f(v,t)$. Then for a *dilute* set of dislocations, changes in the distribution will be described by a Boltzmann-like relaxation equation:

$$\frac{\partial f}{\partial t} = -\left(\frac{f - f_0}{\tau}\right) \tag{8.41}$$

where f_0 is the "free" distribution. This equation indicates that regardless of the initial velocity distribution, the free distribution f_0 is approached in a characteristic time τ. For a *concentrated* set of interacting dislocations, the relaxation equation becomes

$$\frac{\partial f}{\partial t} = -\left(\frac{f - f_0}{\tau}\right) + \left(\frac{\partial f}{\partial t}\right)_I \tag{8.42}$$

where $(\partial f / \partial t)_{\mathrm{I}}$ describes changes in the distribution caused by interactions.

For a simple interaction law that causes both colliding dislocations to acquire the lower of the two velocities possessed by them, the relaxation equation becomes (with $C =$ concentration and $\beta =$ collision cross section)

$$\frac{\partial f}{\partial t} = -\left(\frac{f - f_0}{\tau}\right) + C\beta f \left[1 - 2\int_0^v f(v')\, dv'\right] \tag{8.43}$$

The time-independent solution of this is given by the solution of the following integral equation:

$$f = f_0 + C\beta \tau f \left[1 - 2\int_0^v f(v')\, dv'\right] \tag{8.44}$$

For $v = 0$, the integral in brackets becomes zero, and

$$f = \frac{f_0}{1 - C\beta\tau} \tag{8.45}$$

Therefore $C\beta\tau < 1$, and a critical concentration is reached when

$$C_{\mathrm{crit}} = \frac{1}{\beta\tau} \tag{8.46}$$

At this concentration, a discontinuous *phase transition* occurs and the average velocity drops almost to zero because of intense scattering interactions.

Explicit solutions of the integral equation will not be discussed here, but they have the following general form:

$$f(v) = \frac{f_0(v)}{[1 + C\beta\tau g(v)]^{\frac{1}{2}}} \tag{8.47}$$

Therefore, the average velocity always decreases as the concentration builds up, and then it drops off precipitously when the critical concentration is reached.

8.11 EFFECTS OF HETEROGENEITY

Real materials are seldom homogeneous in terms of composition, structure, or internal stress distribution. Some observers of the subject are dismayed and discouraged by this, because it means that a detailed theory of flow resistance is virtually impossible to formulate. However, it is important to realize that no one really wants a detailed theory anyway, because it would require such a gigantic array of initial conditions that no one would be interested in writing them down. Therefore, it is both desirable and honorable to use the material itself as an analog computer to perform the necessary averaging procedures and give suitable values for the various property parameters. Then the theory can be used to predict response for a new set of conditions.

In order to predict future response from past behavior with confidence, some knowledge of general trends is needed, so some of the effects of heterogeneities will be considered, but no detailed treatment will be attempted.

AMPLIFICATION OF VISCOUS RESISTANCE BY INTERNAL STRESSES

In addition to the complications caused by the heterogeneous nature of real materials, the internal stress fluctuations that are usually present in them can have important effects, especially for nonlinear flow.

For linear viscosity, if the internal stresses fluctuate about some mean value σ with an average amplitude $\Delta\sigma$ and an average wavelength λ, it can be readily shown that the effective damping constant is

$$B'_{eff} = B_0 \left[1 - \left(\frac{\Delta\sigma}{\sigma} \right)^2 \right]^{-1} \tag{8.48}$$

so for small fluctuations there is a negligible effect.

For the nonlinear case, since $e^{1/1+\delta}$ is small compared with $e^{1/1-\delta}$ for moderate values of δ, the effective damping constant has the approximate value

$$B''_{eff} \cong B_{eff} e^{(\sigma/\sigma - \Delta\sigma)} \tag{8.49}$$

so the effective viscosity increases very rapidly with increasing stress-fluctuation amplitude. A more detailed discussion of this effect was first given by Chen, Gilman, and Head (1964), and more recently, it has been considered by Li (1968).

Suppose that a moving dislocation encounters sinusoidally varying internal stresses of amplitude $\Delta\sigma$ and wavelength d (longer than atomic dimensions and shorter than the glide-plane length). Then, as it glides in the x direction on the x-z plane under an applied shear stress σ_0, the net stress encountered is

$$\sigma_{xy} = \sigma_0 + \Delta\sigma \sin \frac{2\pi x}{d} \tag{8.50}$$

The time required to move 1 wavelength is

$$t = \int_0^d \frac{dx}{v(x)} \tag{8.51}$$

where $v(x)$ is given by Eq. (8.18) and the average velocity $\bar{v}$ is given by d/t. If the velocity when $\Delta\sigma = 0$ is v_0 and the reduced drag stress is $\delta = D/\sigma_0$ and the reduced stress-fluctuation amplitude is $\Sigma = \Delta\sigma/\sigma_0$, then the average velocity (after changing the integration variable) is

$$\bar{v} = v_0 \left\{ \int_0^1 \exp\left[\delta \left(1 + \Sigma \sin 2\pi\theta \right)^{-1} - \delta \right] d\theta \right\}^{-1} \tag{8.52}$$

Numerical values for this integral are displayed in Fig. 8.12, and it may be seen that for large drag stresses, the average velocity decreases very rapidly

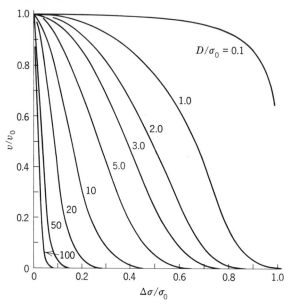

Fig. 8.12   Effect of stress fluctuations on mean dislocation velocities: v_0 = velocity with no fluctuations present; $\bar{v}$ = mean velocity when fluctuations are present; σ_0 = applied stress; $\Delta\sigma$ = fluctuation amplitude: D = characteristic drag stress.

with increasing stress-fluctuation amplitude. Thus the one-dimensional model clearly shows the effect, although real cases are three-dimensional and, hence, more complex.

THE LOCAL STRESS

A severe difficulty associated with any interpretation of flow stresses is that the applied stress may be markedly different from the local stress at a bound state inside a crystal. The local stress may be magnified because of stress relaxation adjacent to it, and it may fluctuate. If the stress dependence of the escape rate were linear, this would not cause great difficulty, because average values of the internal variables would then be useful parameters. However, as shown by Chen, Gilman, and Head (1964), the effects of stress fluctuations are highly nonlinear in typical structural materials. This has also been emphasized by Li (1968) and by Argon (1968).

Consider a very dilute system with a low dislocation density and with high dislocation mobility between pinning points. Then the local stress can be approximated by

$$\sigma_s = \sigma_s^A (C^{-\frac{1}{2}}) \tag{8.53}$$

where σ_s^A is the applied stress and C is the fractional concentration of pinning points. Thus if C is 100 ppm, the local stress is 100 times the applied stress.

At the other extreme when C becomes unity, the local stress equals the applied stress. The intermediate region is the difficult one, because long-range stress fluctuations are then present. They are difficult to define, and as mentioned above, they have nonlinear effects.

MIXTURES OF CHEMICAL PHASES

One of the most important ways to get a fine blend of engineering properties is to use materials that consist of mixtures of hard and soft phases. The best of such materials are both very strong and very tough. In three dimensions the microbehavior in these materials is complex, because dislocation lines in the soft phase can loop around particles of the hard phase (Fig. 4.26) or cross glide and thereby move over them. A complete discussion of the complexities is not feasible here, but many features of them have been reviewed not long ago by Kelly (1963), and the statistics of the configurations were reviewed by Kocks (1969). These reviews are concerned primarily with static models which are not adequate in general, although they may fit limited sets of data. On the other hand, dynamical models give a more physical approach but are statistically complex. All that will be done here is to illustrate the principles using the one-dimensional case.

Variations in the drag stress with position will occur in materials having multiphase structures, banded distributions of dislocations, or clusters of point defects. Most real cases are complex compared to the one-dimensional case. This models a crystal of extent x_1 containing a hard thin zone of extent x_2. If the drag stress in the crystal is D_1 and it is D_2 in the zone (where $D_2 > D_1$), then the time needed for a dislocation to transverse the system is

$$\Delta t = \frac{x_1 v_2 + x_2 v_1}{v_1 v_2} \tag{8.54}$$

and if expressions for v_1 and v_2 are substituted, the average traversal velocity $\bar{v} = \Delta x/\Delta t$ is

$$\bar{v} = \frac{v_1}{1 + (x_2/x_1)(v_1^*/v_2^*)\exp\left[-(D_1 - D_2)/\sigma_s\right]} \tag{8.55}$$

From this it may be seen that even if the ratio x_2/x_1 is small, a significant difference between the two drag stresses will make $\bar{v}$ much less than v_1. This is in good accord with observations. This discussion applies to the motion of an isolated dislocation. In real cases, many dislocations will move in region one, relaxing the stress there and concentrating it in region two. The maximum stress-concentration factor will be approximately $(x_1/x_2)^{\frac{1}{2}}$, so the average drag stress will no longer be $(D_1 - D_2)$, but will become

$$D_1 - \left(\frac{x_1}{x_2}\right)^{\frac{1}{2}} D_2$$

and the effect of the hard layer is thereby reduced by the occurrence of some plastic strain.

In an alloy that is hardened by coherent ordered particles, dislocations move as strongly coupled pairs in the ordered phase because *antiphase boundary* (APB) lies between them. When they move out of the ordered phase, they tend to remain as pairs. Each member of a pair may further split into partials, but this is neglected in the discussion that follows. Because of the necessity for creating APB when a dislocation pair enters the ordered phase, a resistive force arises which delays the entry. Copley and Kear (1967) have investigated the dynamics of this situation in some detail.

The schematic kinematics of the motion of a dislocation pair through an ordered particle is shown in Fig. 8.13, where the ordered particle is designated ω, and the surrounding matrix is μ.

At t_1 the leading member of the pair reaches the μ-ω interface. The leading

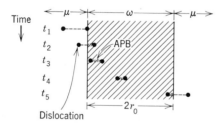

Fig. 8.13 Sequence of positions for a dislocation as it passes through an ordered precipitate particle.

member makes some APB and enters the ω phase at t_2 while the trailing member lags behind in the μ phase. The pair is completely inside ω at t_3 and has reached its steady-state width at t_4. The leading member of the pair leaves ω to enter μ at t_5.

The dynamics of this process was studied by means of the following dislocation-velocity–stress equation:

$$v = v^* \exp - \frac{D}{\sigma_s} = v^* \exp - \frac{F}{\Sigma f_i} \tag{8.56}$$

where the drag stress D and the applied stress σ_s have been replaced by the drag force F, and the sum of other forces f_i caused by the applied stress, the APB, the second member of a pair, etc. Thus, if the leading member of a pair is designated by (1) and the trailing member by (2), their velocities at t_i may be written as follows:

$$
\begin{aligned}
v_1 &= v_\omega^* \exp \left[\frac{-F_\omega}{\sigma_s b - A + K/w + \theta_1(x)} \right] \\
v_2 &= v_\mu^* \exp \left[\frac{-F_\mu}{\sigma_s b + K/w + \theta_3(x)} \right]
\end{aligned}
\tag{8.57}
$$

where F_ω and F_μ are the drag forces in the two phases; $\sigma_s b$ is the stress-induced force; A is the force caused by the APB; K/w (with w being the separation width and K an elastic coefficient) is the force caused by the other dislocation; and θ is the line tension which is a function of the lateral coordinate x. The velocities and displacements were determined by means of stepwise integration of these functions.

It was found that a dislocation spends most of its time in the process that is in progress at t_2 of Fig. 8.13. Hence an approximate expression for the average velocity can be written

$$\bar{v} = \exp \left[\frac{-(F_\omega + F_\mu)}{2\sigma_s b - A + Gb^2/r_0} \right] \tag{8.58}$$

where Gb/r_0 is the effect of the line tension, with r_0 being the particle radius and G the shear modulus.

The results of numerical computations for a nickel-base alloy (MAR M-200) yielded the following average velocities:

Velocity in pure matrix $= 1.7 \times 10^3$ cm/sec
Velocity in ordered phase $= 3.3 \times 10^2$ cm/sec
Velocity in mixture $= 10^{-3}$ cm/sec

which shows the large effect of the time spent in entering the ordered phase on the average velocity. A comparison between the dynamical calculations and the observed behavior of the alloy is shown in Fig. 8.14, where it may be seen that quite reasonable agreement is obtained.

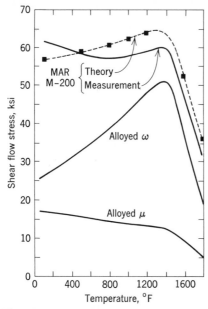

Fig. 8.14   Comparison of observed flow stresses of a nickel-base alloy (MAR M-200) and calculation of dynamical behavior. (*After Copley and Kear.*)

8.12 FLOW CATALYSIS

In covalent crystals, the evidence is clear that dislocation motion involves "bond-breaking" and therefore may be considered to be a special kind of chemical reaction. Just as reactions can be affected by auxiliary influences, it might be expected that flow could be, and indeed it is. Many of the effects are localized at free surfaces, and these have been summarized in a recent book edited by Westwood (1968). In at least one case, however, an action that is closely analogous to catalysis is observed at the interior of crystals.

Dry quartz crystals have compressive strengths of about 20 kbar at room temperature and retain approximately this strength up to 1000°C. However, the presence of small amounts of water causes softening above about 800°C (Griggs and Blacic, 1965). It has been postulated that water facilitates the flow by hydrolyzing the places along dislocations where Si-O bridges are broken. The resulting silanol groups have enough mobility at high temperatures to move along with a dislocation and repetitively aid in the shear breakage of bonds. Thus they act as flow catalysts.

Water also affects the mobilities of dislocations near the surfaces of several crystals such as MgO, Al_2O_3, Ge, and TiC (Westbrook and Jorgensen, 1965). Other solvents such as dimethyl formamide may also have large effects (Westwood, 1968), and the effect of water can be markedly changed by adding salts such as silver chloride to it.

8.13 PLASTIC ANISOTROPY

An important distinction between the flow of a glass and that of a crystal is that the later may be quite anisotropic. There are several forms of this anisotropy which are related to the symmetry properties of crystals.

PLANES

It was mentioned previously (Sec. 3.2) that a given glide direction may be associated with more than one crystallographic glide plane and that their flow stresses may be quite different in magnitude. One example is provided by beryllium, as studied by Levine, Kaufman, and Aronin (1964), and shown in Fig. 8.15. Similar results have been obtained for Zn and Cd by the author, as well as for LiF (Gilman, 1959).

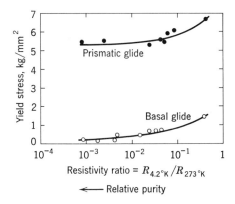

Fig. 8.15 Stress required for basal and prismatic glide in beryllium as a function of purity. (*After Levine, Kaufman, and Aronin.*)

A particularly interesting case is crystals which have the zinc blende structure. In them, the structures of the cores of positive edge dislocations are different from those of negative edge dislocations. In one case the extra half plane ends on an A-type atom, whereas it ends on B-type atoms in the other case. This leads to different mobilities for the two dislocation types. For example, in InSb, the type that ends on In is more mobile than the Sb type (Bell and Willoughby, 1966).

In some cases the preference for motion on one plane compared with another is rather small. This is the case for iron and some iron alloys near room temperature. At low temperatures (or high strain rates) the $\{110\}$ planes are preferred, but at higher temperatures (or lower strain rates) glide tends to take place along the plane that has the maximum shear stress on it. It does this by means by a series of zigzag motions on a fine scale, where the ratio of zig to zag determines the average plane of the motion. The behavior is complicated by the fact that the rates of motion on the component planes depend on the sense of the applied shear stress. Thus compression in a given crystallographic direction yields somewhat different behavior than does tension in the same direction. This is also observed in other body-centered-cubic metals, such as tantalum and niobium. The theory of this "composite glide" has been given by Vitek and Kroupa (1966) in terms of thermally activated splitting of screw dislocations onto different planes. This splitting vanishes and reappears repetitively. The ratio of the frequency with which it occurs for one plane compared with another determines the orientation of the average glide plane.

STRESS DIRECTIONS

Depending on what the symmetry of a crystal is, its glide behavior may be quite different for shear in one direction than for the opposite one. This is aptly called unidirectional glide. It is related to twinning where only one sense of the applied shear is able to cause the process to occur for a reasonable magnitude of the stress.

REFERENCES

Alexander, H., and P. Haasen: *Solid State Phys.*, **22**: 27 (1968).

Andrade, E. N. da C.: *Phil. Mag.*, **17**: 497, 698 (1934).

Argon, AS.: Dislocation Dynamics, *Mater. Sci. Eng.*, **3**: 24 (1968).

Bell, R. L., and A. F. W. Willoughby: *J. Mater. Sci.*, **1**: 219 (1966).

Blish, R. C., and T. Vreeland: *J. Appl. Phys.*, **39**: 2816 (1968).

Chen, H. S., J. J. Gilman, and A. K. Head: Dislocation Multipoles and their Role in Strain-hardening, *J. Appl. Phys.*, **35**: 2502 (1964).

Copley, S. M., and B. H. Kear: *Trans. AIME*, **239**: 984 (1967).

Cottrell, A. H.: "Dislocations and Plastic Flow in Crystals," pp. 56, 142, 152, Oxford University Press, Fair Lawn, N.J., 1953.

Dash, W. C.: *J. Appl. Phys.*, **29**: 705 (1958).

Eshelby, J. D.: The Interaction of Kinks and Elastic Waves, *Proc. Roy. Soc.*, **266A**: 222 (1962).

Farren, W. S., and G. I. Taylor: The Heat Developed during Plastic Extension of Metals, *Proc. Roy. Soc.*, **107**: 422 (1925).

Fourie, J. T., and R. J. Murphy: *Phil. Mag.*, **7**: 1617 (1962).

Frenkel, J.: *Z. Physik.*, **35**: 652 (1926).

Gilman, J. J.: *Proceedings of the Battelle Conference on Dislocation Dynamics*, McGraw-Hill Book Company, New York, 1968.

———: *J. Met.*, **18**: 1171 (1966); also, *Bull. Am. Phys. Soc.*, **12**: 370 (1967); also, *Proceedings of the Battelle Conference on Dislocation Dynamics*, McGraw-Hill Book Company, New York, 1968.

———: Dislocation Motion in a Viscous Medium, *Phys. Rev. Letters*, **20**: 157 (1968).

———: Dislocation Mobility in Crystals, *J. Appl. Phys.*, **36**: 3195 (1965).

———: *Australian J. Phys.*, **13**: 327 (1960).

———: *Acta Met.*, **7**: 608 (1959).

Glen, J. W.: *Phil. Mag.*, **1**: 400 (1956).

Green, H. S.: "The Molecular Theory of Fluids," Interscience Publishers, New York, 1952.

Greenman, W. F., T. Vreeland, and D. S. Wood: Dislocation Mobility in Copper, *J. Appl. Phys.*, **38**: 3595 (1967).

Griggs, D. T., and J. D. Blacic: *Science*, **147**: 292 (1965).

Johnston, W. G.: *J. Appl. Phys.*, **33**: 2050 (1962).

——— and J. J. Gilman: *J. Appl. Phys.*, **31**: 632 (1960).

Kelly, A.: in "Electron Microscopy and Strength of Crystals," p. 947, Interscience Publishers, a division of John Wiley & Sons, Inc., New York, 1963.

Kocks, U. F.: in "Physics of Strength and Plasticity," Orowan Anniv. Vol., A. S. Argon (ed.), The M.I.T. Press, Cambridge, Mass., 1969.

Kontorova, T. A., and Y. I. Frenkel: *J. Exp. Theor. Phys.* (U.S.S.R.), **8**: 89, 1340 (1938).

Kratochvil, J.: Atomic Models of Crystal Defects, in "Theory of Crystal Defects," p. 17, Academic Press Inc., New York, 1966.

———— and V. L. Indenbom: The Mobility of a Dislocation in the Frenkel-Kontorova Model, *Czech. J. Phys.*, **13**: 814 (1963).

Kuhlmann-Wilsdorf, D.: *Phys. Rev.*, **120**: 773 (1960).

Levine, E. D., D. F. Kaufman, and L. R. Aronin: *Trans. AIME*, **230**: 260 (1964).

Li, J. C. M.: *Proceedings of the Battelle Conference on Dislocation Dynamics*, McGraw-Hill Book Company, New York, 1968.

Mason, W. P.: "Physical Acoustics," vol. 3, part B, and vol. 4, part A, Academic Press Inc., New York, 1965.

————: Phonon Viscosity and Its Effect on Acoustic Wave Attenuation and Dislocation Motion, *J. Acoust. Soc. Am.*, **32**: 458 (1960).

Maxwell, J. C.: *Phil. Trans. Roy. Soc.* (*London*), **157A**: 49 (1867).

Mott, N. F.: *Phil. Mag.*, **1**: 568 (1956).

Nabarro, F. R. N.: "Theory of Crystal Dislocations," p. 693ff, Oxford University Press, Fair Lawn, N.J., 1967.

Nikitenko, B. I.: "Dislocations and Physical Properties of Semiconductors," A. R. Regel (ed.), p. 58, Acad. Sci. U.S.S.R., Moscow, 1967.

Pope, D. P., T. Vreeland, Jr., and D. S. Wood: Mobility of Edge Dislocations in the Basal-Ship System of Zinc, *J. Appl. Phys.*, **38**: 4011 (1967).

Prigogine, I.: "Theory of Traffic Flow," p. 158, Elsevier Publishing Company, New York (1961).

Sanders, W. T.: *J. Appl. Phys.*, **36**: 2822 (1965).

Schafer, S.: *Phys. Stat. Solidi*, **19**: 297 (1967).

Schoeck, G., and A. Seeger: The Flow Stress of Iron and Its Dependence on Impurities, *Acta Met.*, **7**: 469 (1959).

Suzuki, H.: A Note on the Peierls Force, *Proceedings of the International Conference on Crystal Lattice Definition*, *J. Phys. Soc. Jap.*, **18**: suppl. 1, p. 182 (1963).

Takamura, J., and T. Morimoto: Dynamic Interaction between a Moving Dislocation and Point Defects, *J. Phys. Soc. Jap.*, **18**: suppl. 16, p. 28 (1963).

Taylor, J. W.: "Hypervelocity Impact Phenomena," E. Cohen (ed.), Academic Press Inc., New York (1969).

Vitek, V., and F. Kroupa: *Phys. Stat. Solidi*, **18**: 703 (1966).

Weertman, J.: Stress Dependence of the Velocity of a Dislocation Moving on a Viscously Damped Slip Plane, in "Physics of Strength and Plasticity," A. Argon (ed.), The M.I.T. Press, Cambridge, Mass., 1969.

————: *J. Appl. Phys.*, **29**: 1685 (1958).

Westbrook, J. H., and P. J. Jorgensen: *Trans. AIME*, **230**: 425 (1965).

Westwood, A. R. C.: in "Microplasticity," C. J. McMahon (ed.), p. 365, Interscience Publishers, a division of John Wiley & Sons, Inc., New York, 1968.

Zener, C.: *Proc. Roy. Soc. (London)*, **145**: 523 (1934).

INDEX

This book was set in Modern by The Maple Press Company, and printed on permanent paper and bound by The Maple Press Company. The designer was Marsha Cohen; the drawings were done by J. & R. Technical Services, Inc. The editors were Bradford Bayne and Maureen McMahon. Sally R. Ellyson supervised the production.